student study
ART NOTEBOOK

Biology

Sixth Edition

John W. Kimball

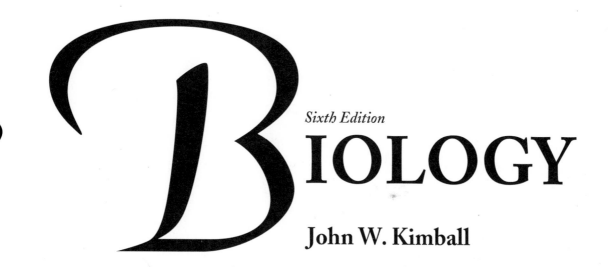

WCB **Wm. C. Brown Publishers**

Dubuque, Iowa · Melbourne, Australia · Oxford, England

Book Team

Editor *Carol J. Mills*
Developmental Editor *Diane Beausoleil*
Production Editor *Catherine S. Di Pasquale*
Designer *Anna C. Manhart*
Art Editor *Joseph P. O'Connell*
Photo Editor *Carrie Burger*
Permissions Coordinator *Mavis M. Oeth*

 Wm. C. Brown Publishers
A Division of Wm. C. Brown Communications, Inc.

Vice President and General Manager *Beverly Kolz*
Vice President, Publisher *Kevin Kane*
Vice President, Director of Sales and Marketing *Virginia S. Moffat*
National Sales Manager *Douglas J. DiNardo*
Marketing Manager *Craig Johnson*
Advertising Manager *Janelle Keeffer*
Director of Production *Colleen A. Yonda*
Publishing Services Manager *Karen J. Slaght*
Permissions/Records Manager *Connie Allendorf*

 Wm. C. Brown Communications, Inc.

President and Chief Executive Officer *G. Franklin Lewis*
Corporate Senior Vice President, President of WCB Manufacturing *Roger Meyer*
Corporate Senior Vice President and Chief Financial Officer *Robert Chesterman*

The credits section for this book begins on page 101 and is considered
an extension of the copyright page.

TO INSTRUCTORS AND STUDENTS

The Student Study Art Notebook is free with a new textbook to all students and can be used to take notes during lectures. On each notebook page, there are two figures (sometimes one, sometimes three) faithfully reproduced from the original textbook figure. Each figure also corresponds to each of the 148 acetates available to instructors with adoption of the text.

The intention is to place a copy of the transparency acetate art in front of students (via the notebook) as the instructor uses the overhead during lectures. The advantage to the student is that he/she will be able to see all labels clearly, and take meaningful notes without having to make hurried sketches of the acetate figure.

The pages of the Art Notebook are perforated and three-hole punched, so they can be removed and placed in a personal binder for specific study and review, or to create space for additional notes.

DIRECTORY OF NOTEBOOK FIGURES

TO ACCOMPANY BIOLOGY, 6/E BY JOHN W. KIMBALL

Composition of Lithosphere		Composition of Human Body	
Oxygen	47	Hydrogen	63
Silicon	28	Oxygen	25.5
Aluminum	7.9	Carbon	9.5
Iron	4.5	Nitrogen	1.4
Calcium	3.5	Calcium	0.31
Sodium	2.5	Phosphorus	0.22
Potassium	2.5	Chlorine	0.03
Magnesium	2.2	Potassium	0.06
Titanium	0.46	Sulfur	0.05
Hydrogen	0.22	Sodium	0.03
Carbon	0.19	Magnesium	0.01
All others	<0.1	All others	<0.01

Elemental composition of the earth's crust and the human body
Figure 1.2

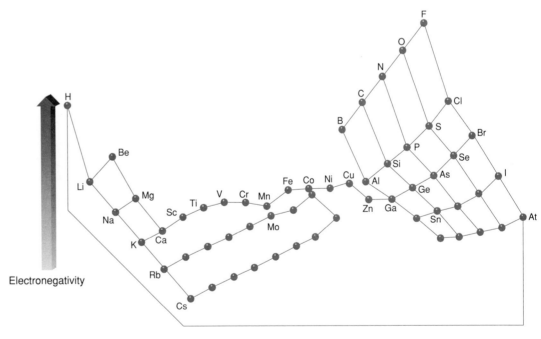

Relative electronegativities
Figure 2.9

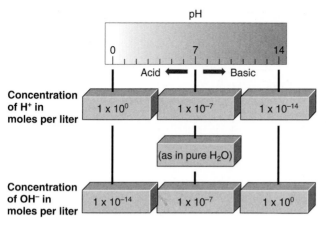

pH= − log (exponent) of the concentration of H^+
in moles per liter

The pH scale
Figure 2.20

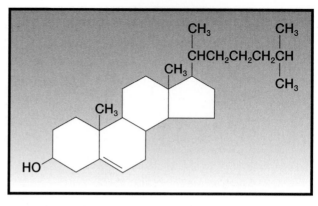

Cholesterol, the most abundant steroid in the human body
Figure 3.9

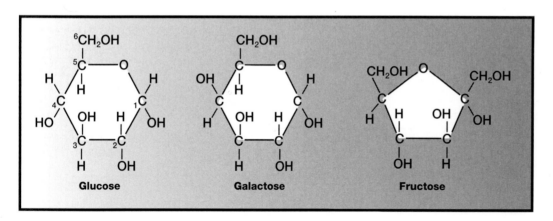

Structural formulas of glucose and two of its isomers
Figure 3.10

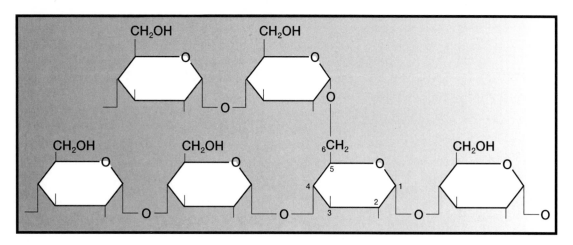

Structure of starch
Figure 3.12

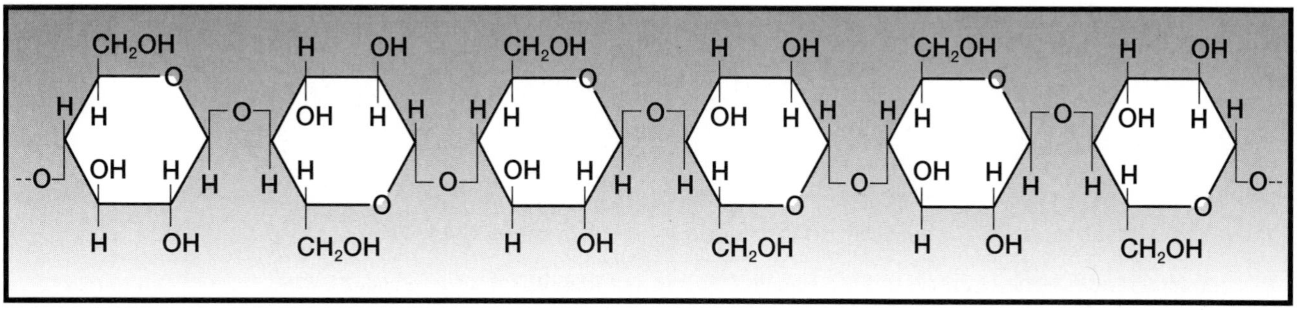

Structure of cellulose
Figure 3.15

3

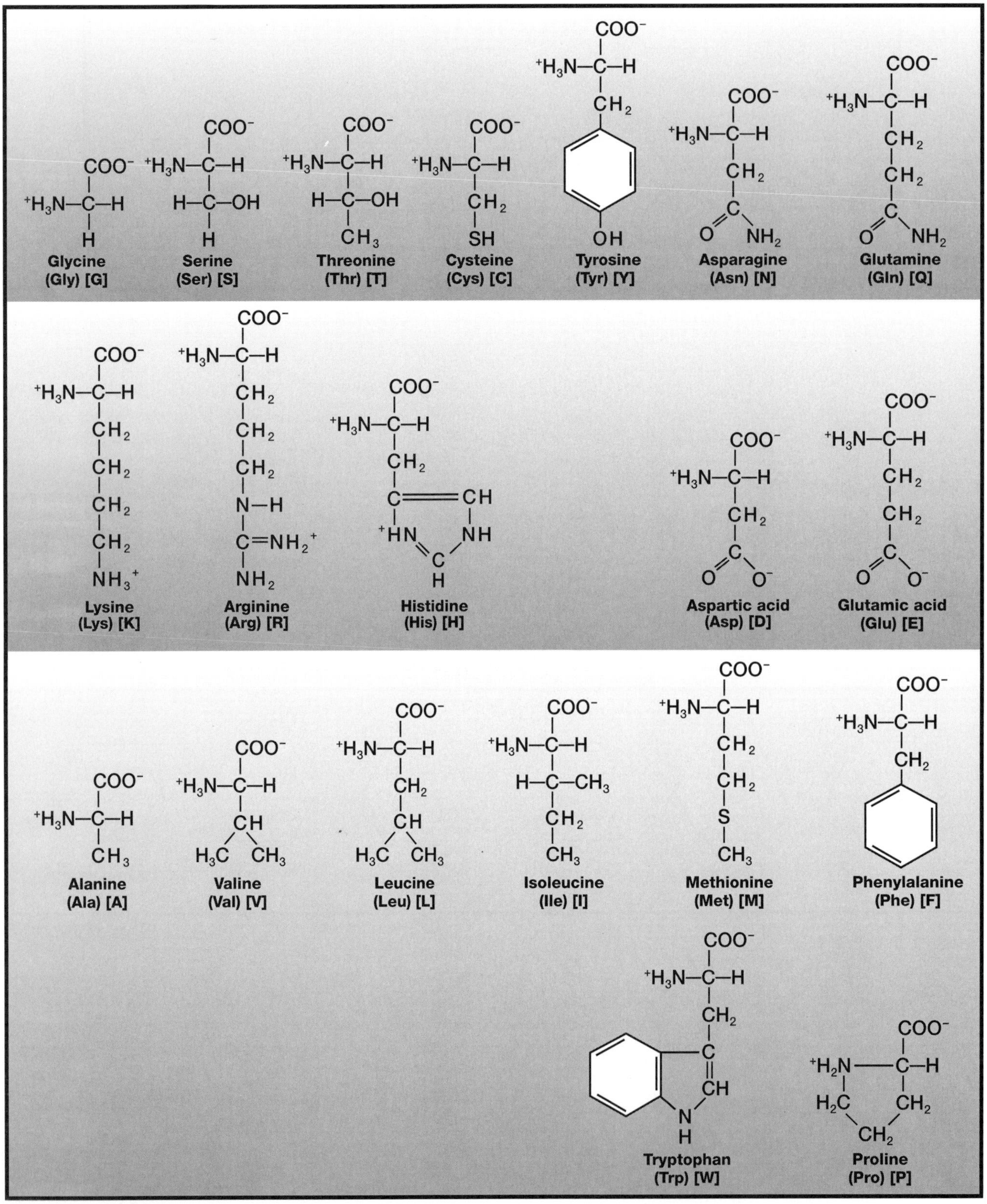

**Structures of the 20 amino acids from which proteins
are synthesized**
Figure 3.17

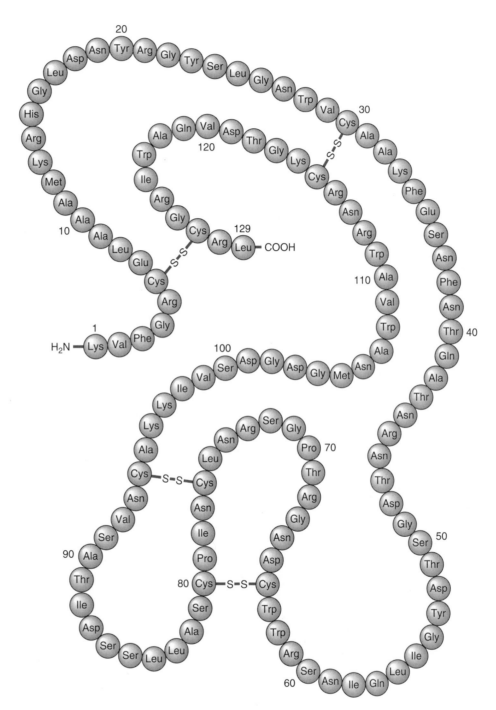

Primary structure of lysozyme
Figure 3.24

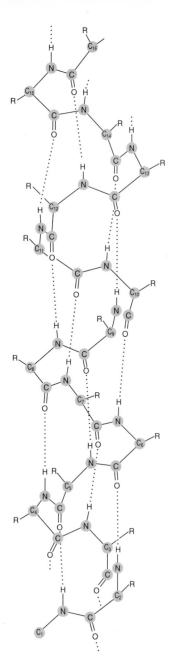

The alpha helix
Figure 3.25

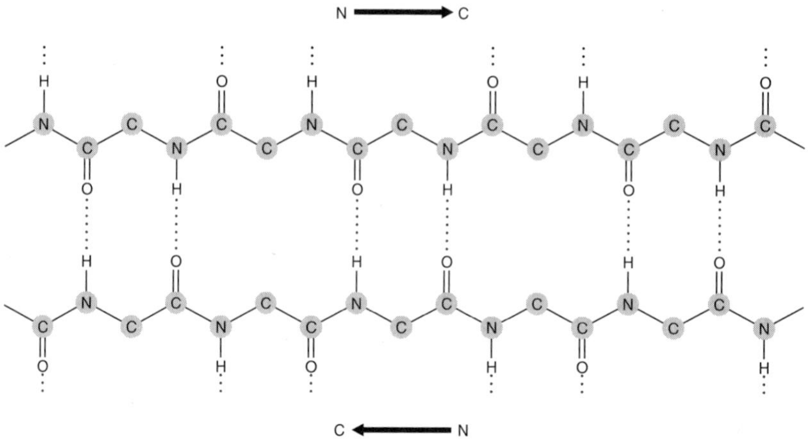

The beta-pleated sheet
Figure 3.26

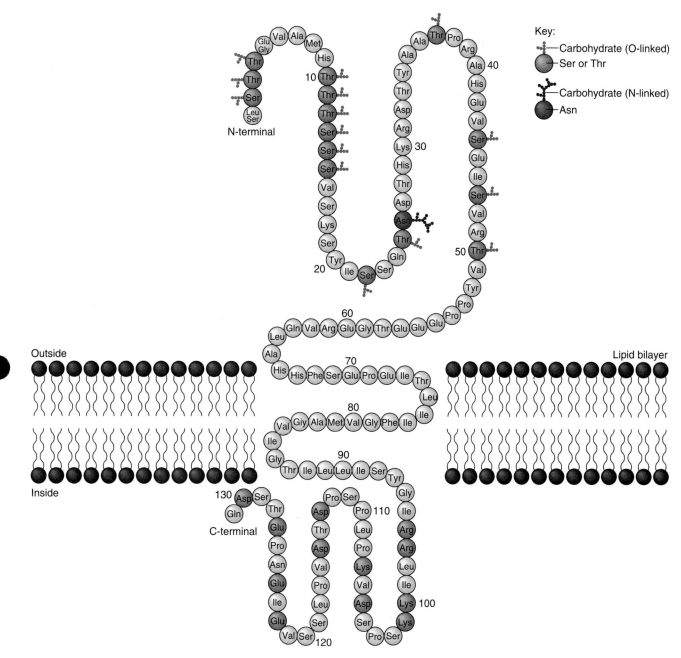

Primary structure of glycophorin A
Figure 3.31

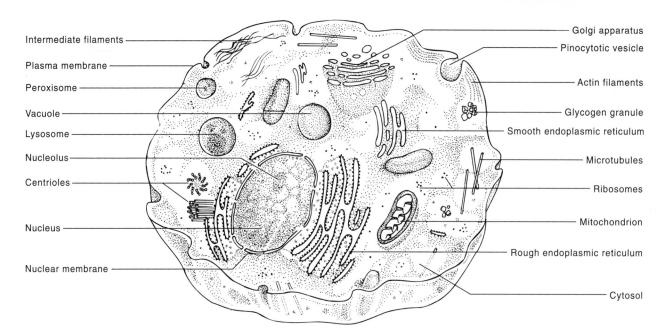

Idealized view of an animal cell as seen under an electron microscope
Figure 4.8

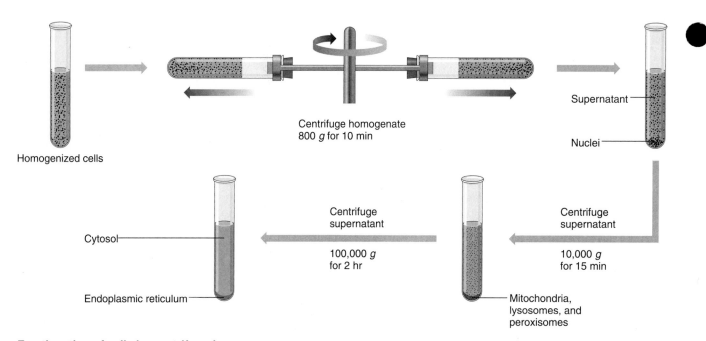

Fractionation of cells by centrifugation
Figure 4.9

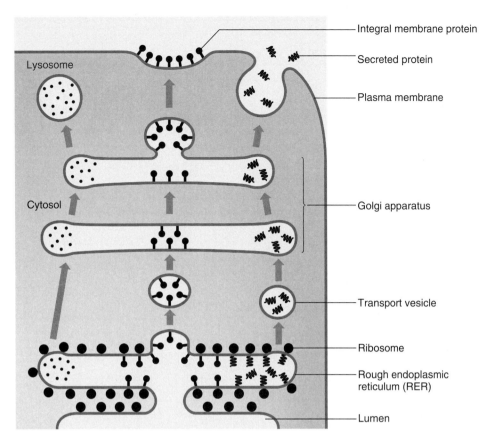

**Transport of proteins from the endoplasmic reticulum
to the Golgi apparatus**
Figure 4.12

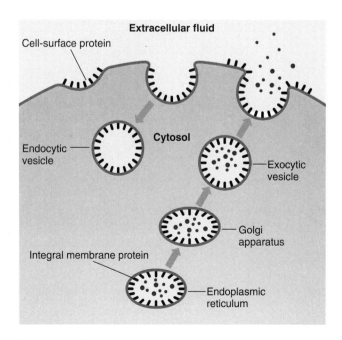

Topology of endocytosis and exocytosis
Figure 4.36

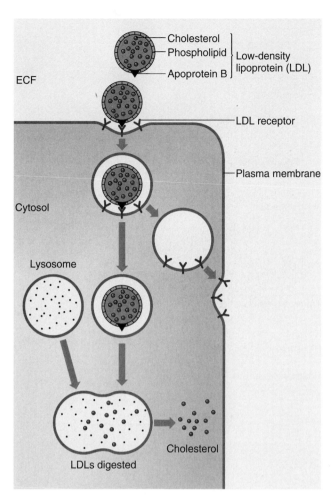

How cells acquire cholesterol
Figure 4.40

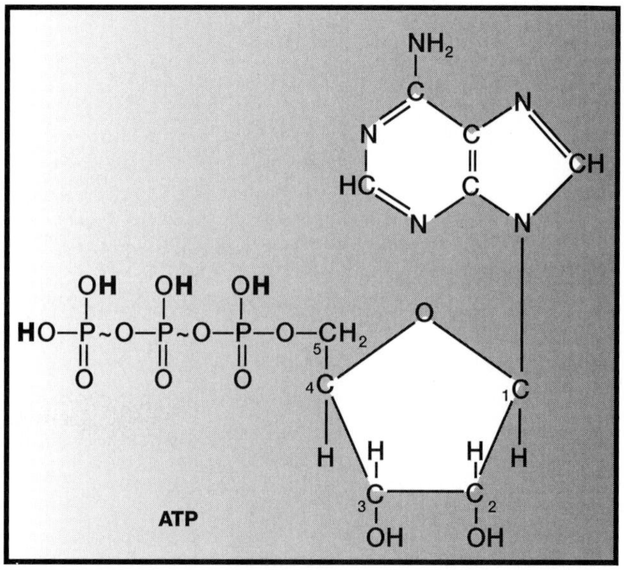

Structure of ATP
Figure 5.1

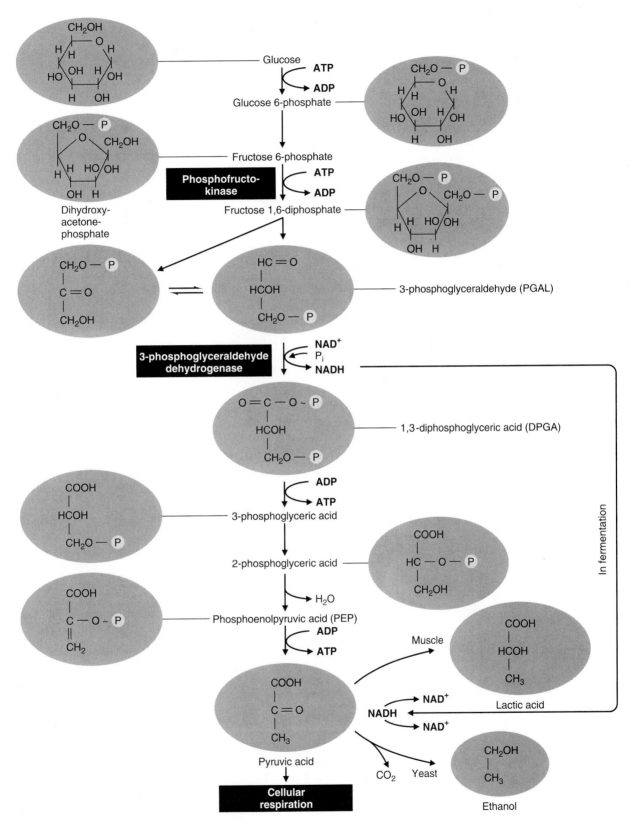

The pathway of glycolysis
Figure 5.3

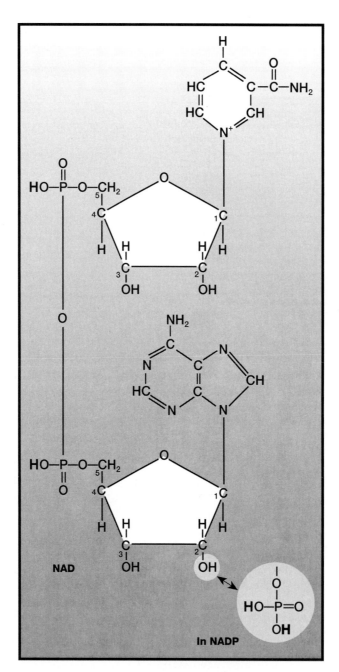

Structures of NAD and NADP
Figure 5.4

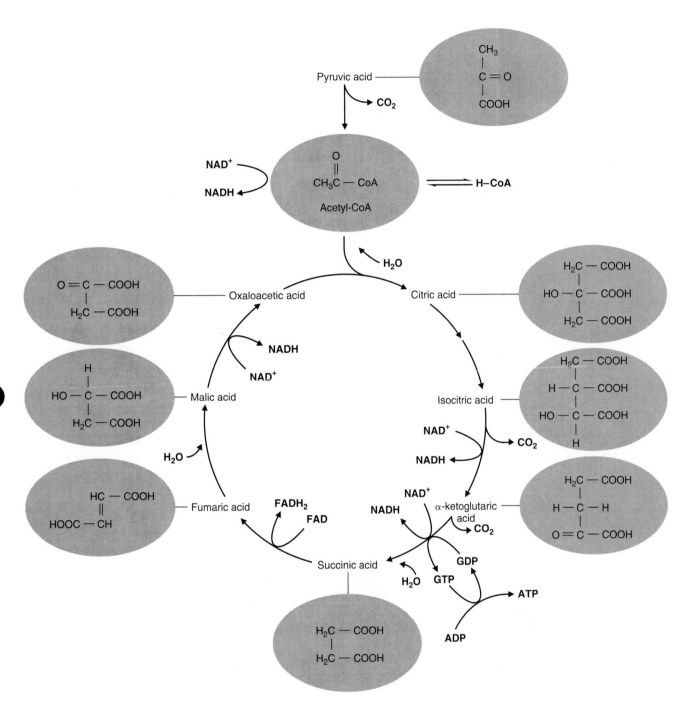

The pathway of cellular respiration
Figure 5.6

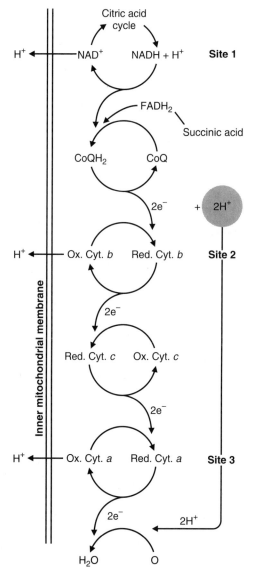

Citric acid cycle

H^+ ← NAD$^+$ NADH + H$^+$ **Site 1**

FADH$_2$

Succinic acid

CoQH$_2$ CoQ

$2e^-$ + $2H^+$

H^+ ← Ox. Cyt. b Red. Cyt. b **Site 2**

$2e^-$

Red. Cyt. c Ox. Cyt. c

$2e^-$

H^+ ← Ox. Cyt. a Red. Cyt. a **Site 3**

$2e^-$ $2H^+$

H_2O O

Inner mitochondrial membrane

(For one glucose molecule, $24e^- + 24H^+ + 6O_2 \longrightarrow 12H_2O$)

The respiratory chain
Figure 5.7

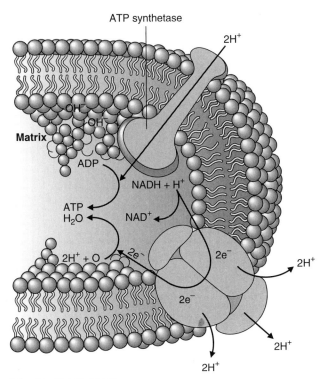

How chemiosmosis couples electron transport to the synthesis of ATP
Figure 5.10

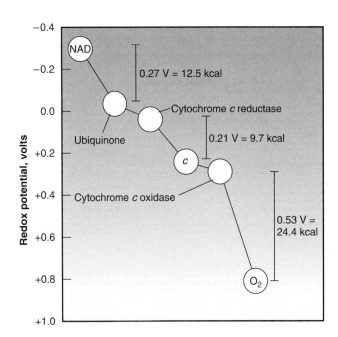

Redox potentials of the components of the respiratory chain
Figure 5.13

15

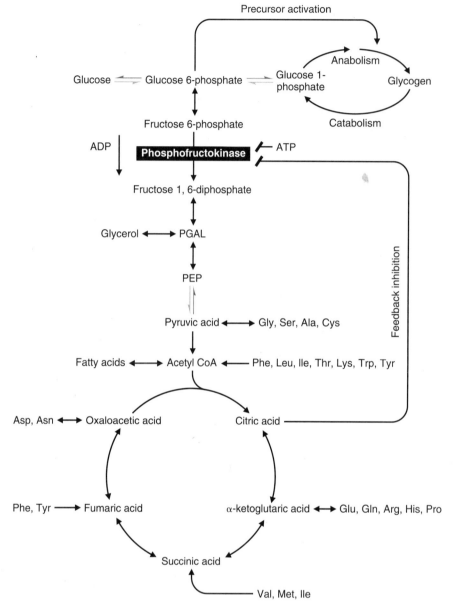

Interconnections between fat, protein, and carbohydrate metabolism
Figure 5.14

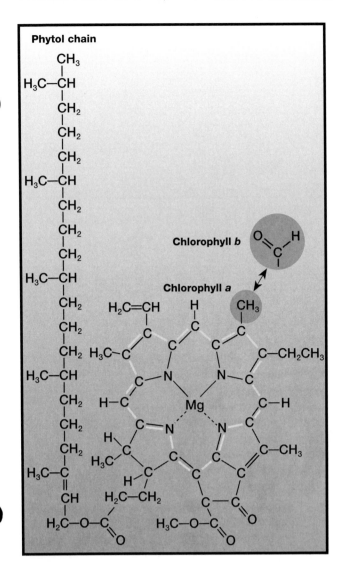

Structures of chlorophyll *a* and *b*
Figure 5.18

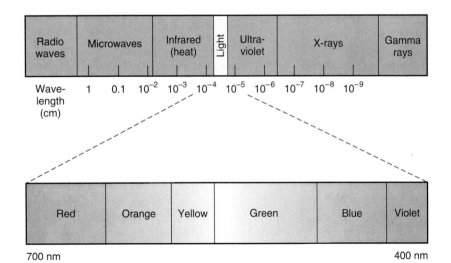

The spectrum of electromagnetic radiation
Figure 5.19

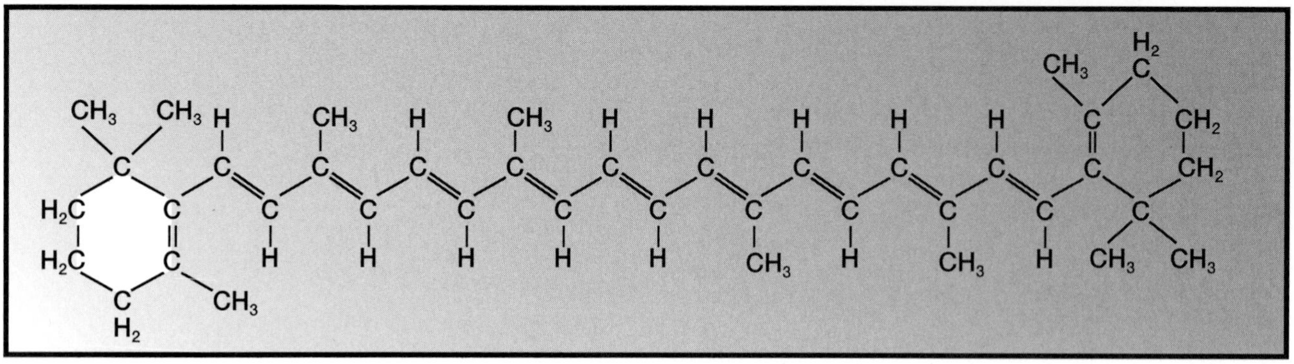

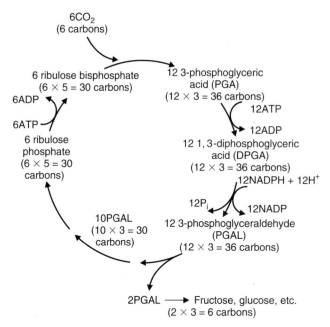

Beta-carotene, one of the most abundant carotenoids
Figure 5.21

6CO$_2$
(6 carbons)

6 ribulose bisphosphate
(6 × 5 = 30 carbons)

12 3-phosphoglyceric
acid (PGA)
(12 × 3 = 36 carbons)

6ADP

6ATP

12ATP

12ADP

6 ribulose
phosphate
(6 × 5 = 30
carbons)

12 1,3-diphosphoglyceric
acid (DPGA)
(12 × 3 = 36 carbons)

12NADPH + 12H$^+$

12P$_i$

12NADP

10PGAL
(10 × 3 = 30
carbons)

12 3-phosphoglyceraldehyde
(PGAL)
(12 × 3 = 36 carbons)

2PGAL ⟶ Fructose, glucose, etc.
(2 × 3 = 6 carbons)

**Pathway of carbon dioxide fixation in photosynthesis—
the "dark" reactions**
Figure 5.27

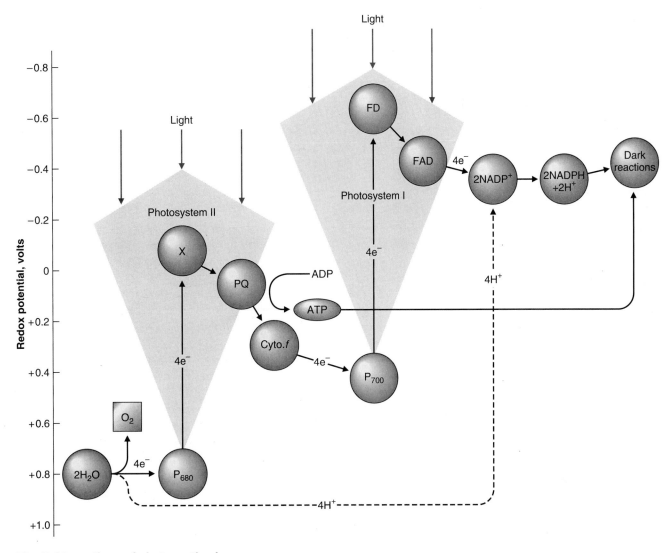

The light reactions of photosynthesis
Figure 5.30

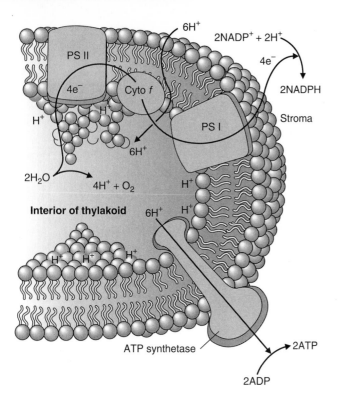

How chemiosmosis couples electron transport in the thylakoid membrane to ATP synthesis
Figure 5.31

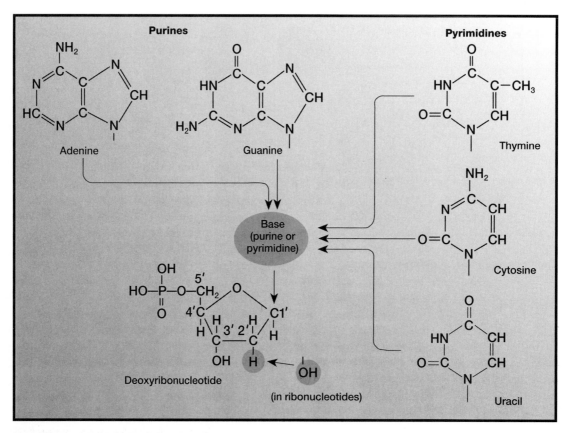

Structures of the nucleotides that serve as the monomers of DNA and RNA
Figure 6.7

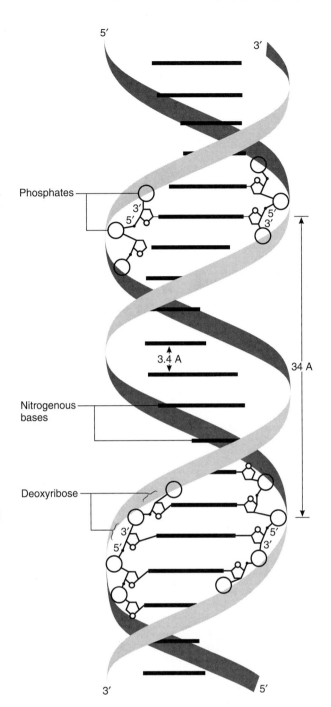

Phosphates

3′
5′

3.4 A

34 A

Nitrogenous
bases

Deoxyribose

3′
5′

3′
5′

5′

3′

The structure of DNA
Figure 6.11

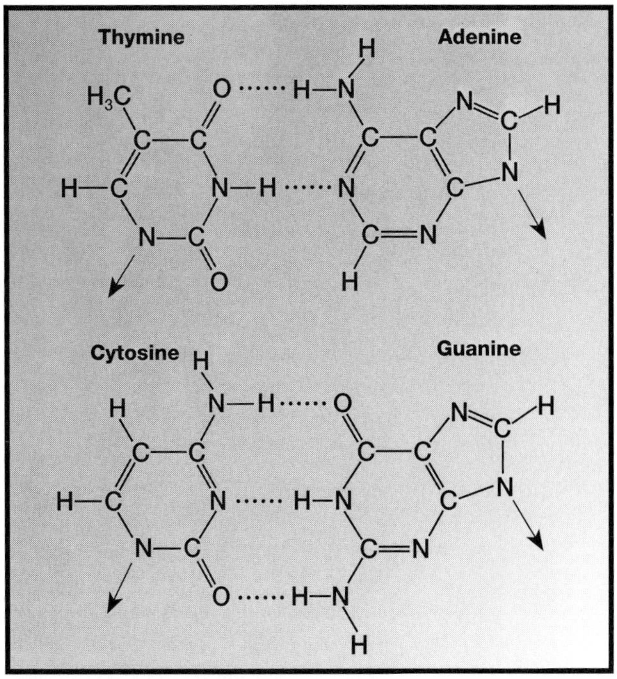

Thymine   Adenine

Cytosine   Guanine

**Location of the hydrogen bonds between thymine (T)
and adenine (A) and between cytosine (C) and guanine
(G)**
Figure 6.12

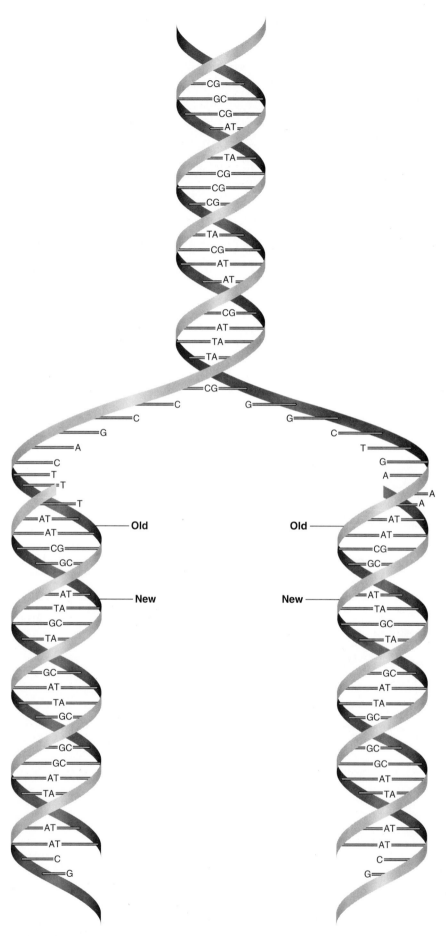

The mechanism of DNA duplication
Figure 6.13

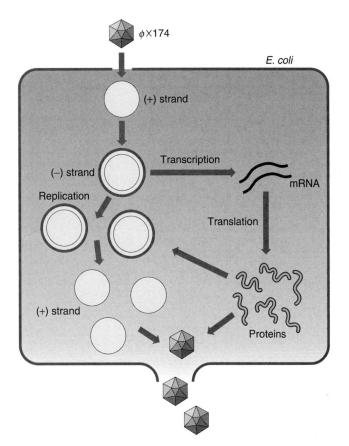

Life cycle of φX174
Figure 6.19

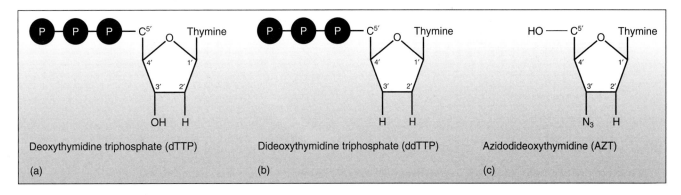

Deoxythymidine triphosphate (dTTP)

(a)

Dideoxythymidine triphosphate (ddTTP)

(b)

Azidodideoxythymidine (AZT)

(c)

Structure of dTTP, ddTTP, and AZT
Figure 6.21

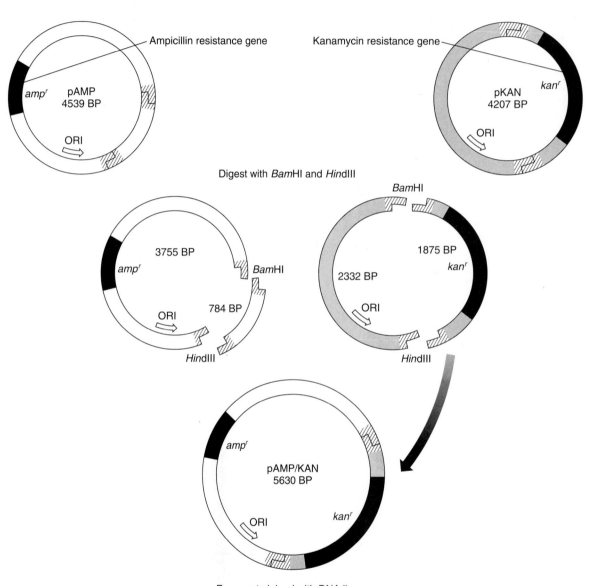

Ampicillin resistance gene

amp^r pAMP
4539 BP

ORI

Kanamycin resistance gene

pKAN
4207 BP kan^r

ORI

Digest with *Bam*HI and *Hind*III

3755 BP

amp^r

ORI 784 BP

*Bam*HI

*Hind*III

*Bam*HI

2332 BP 1875 BP kan^r

ORI

*Hind*III

amp^r

pAMP/KAN
5630 BP

ORI kan^r

Fragments joined with DNA ligase

Making recombinant DNA
Figure 6.26

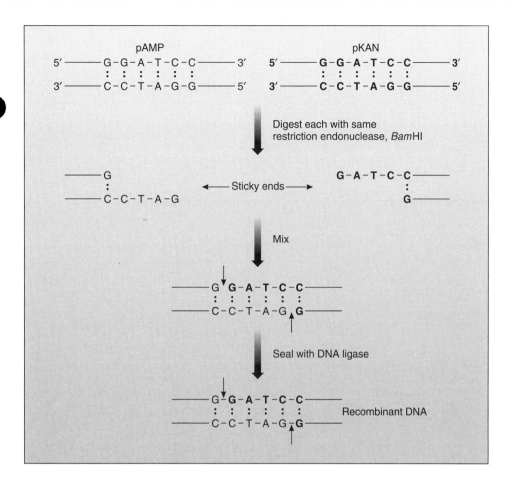

Digestion of pAMP and pKAN by *Bam*HI
Figure 6.28

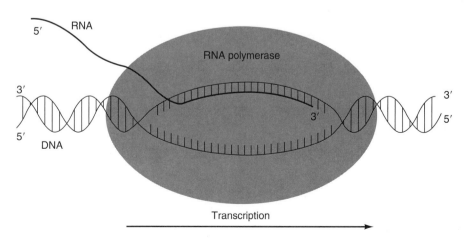

Gene transcription
Figure 7.1

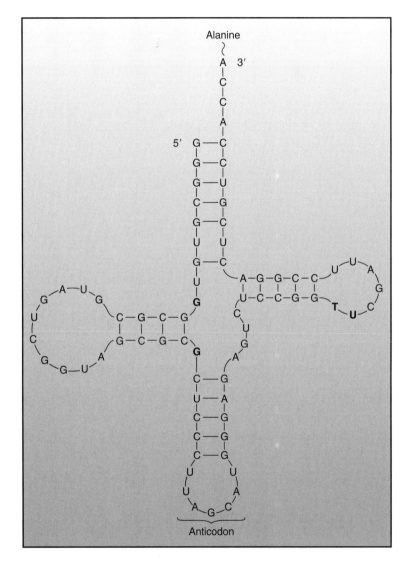

Structure of alanine transfer RNA from yeast
Figure 7.4

Second nucleotide

		U	C	A	G	
First nucleotide	**U**	UUU UUC } Phenylalanine (Phe) UUA UUG } Leucine (Leu)	UCU UCC UCA UCG } Serine (Ser)	UAU UAC } Tyrosine (Tyr) UAA STOP UAG STOP	UGU UGC } Cysteine (Cys) UGA STOP UGG Tryptophan (Trp)	U C A G
	C	CUU CUC CUA CUG } Leucine (Leu)	CCU CCC CCA CCG } Proline (Pro)	CAU CAC } Histidine (His) CAA CAG } Glutamine (Gln)	CGU CGC CGA CGG } Arginine (Arg)	U C A G
	A	AUU AUC } Isoleucine (Ile) AUA AUG Methionine (Met) or START	ACU ACC ACA ACG } Threonine (Thr)	AAU AAC } Asparagine (Asn) AAA AAG } Lysine (Lys)	AGU AGC } Serine (Ser) AGA AGG } Arginine (Arg)	U C A G
	G	GUU GUC GUA GUG } Valine (Val)	GCU GCC GCA GCG } Alanine (Ala)	GAU GAC } Aspartic acid (Asp) GAA GAG } Glutamic acid (Glu)	GGU GGC GGA GGG } Glycine (Gly)	U C A G

Third nucleotide

The genetic code
Figure 7.7

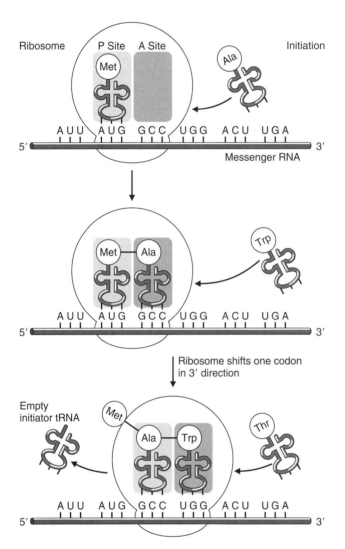

Mechanism of polypeptide synthesis
Figure 7.8

The Genetic Code (DNA)			
TTT = Phe	TCT = Ser	TAT = Tyr	TGT = Cys
TTC = Phe	TCC = Ser	TAC = Tyr	TGC = Cys
TTA = Leu	TCA = Ser	TAA = STOP	TGA = STOP
TTG = Leu	TCG = Ser	TAG = STOP	TGG = Trp
CTT = Leu	CCT = Pro	CAT = His	CGT = Arg
CTC = Leu	CCC = Pro	CAC = His	CGC = Arg
CTA = Leu	CCA = Pro	CAA = Gln	CGA = Arg
CTG = Leu	CCG = Pro	CAG = Gln	CGG = Arg
ATT = Ile	ACT = Thr	AAT = Asn	AGT = Ser
ATC = Ile	ACC = Thr	AAC = Asn	AGC = Ser
ATA = Ile	ACA = Thr	AAA = Lys	AGA = Arg
ATG = Met*	ACG = Thr	AAG = Lys	AGG = Arg
GTT = Val	GCT = Ala	GAT = Asp	GGT = Gly
GTC = Val	GCC = Ala	GAC = Asp	GGC = Gly
GTA = Val	GCA = Ala	GAA = Glu	GGA = Gly
GTG = Val	GCG = Ala	GAG = Glu	GGG = Gly

*When within gene; at beginning of gene, ATG signals START of translation.

The Rosetta Stone of life
Figure 7.10

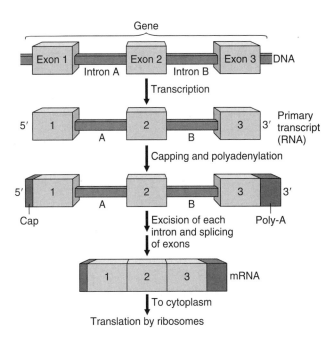

Organization of a split gene
Figure 7.12

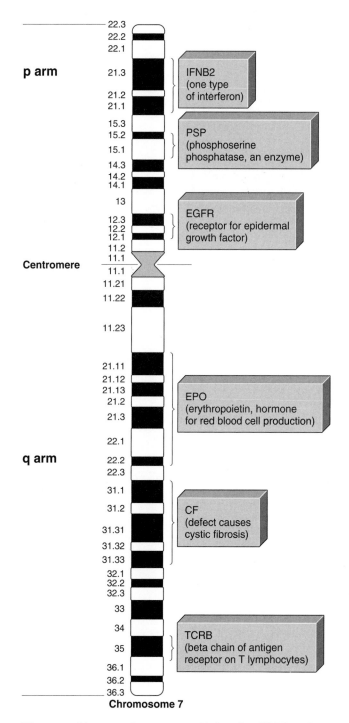

Diagram of human chromosome 7 showing "G" bands
Figure 8.10

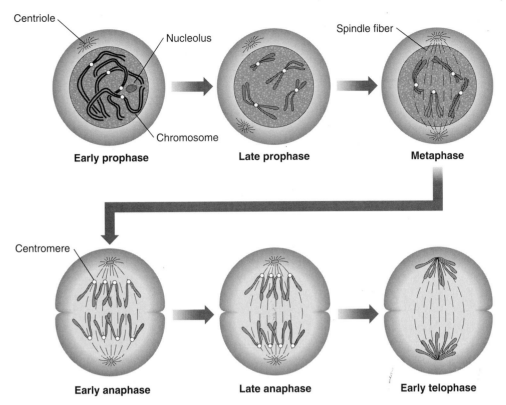

Mitosis of an animal cell
Figure 8.14

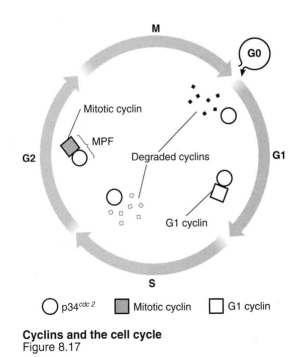

Cyclins and the cell cycle
Figure 8.17

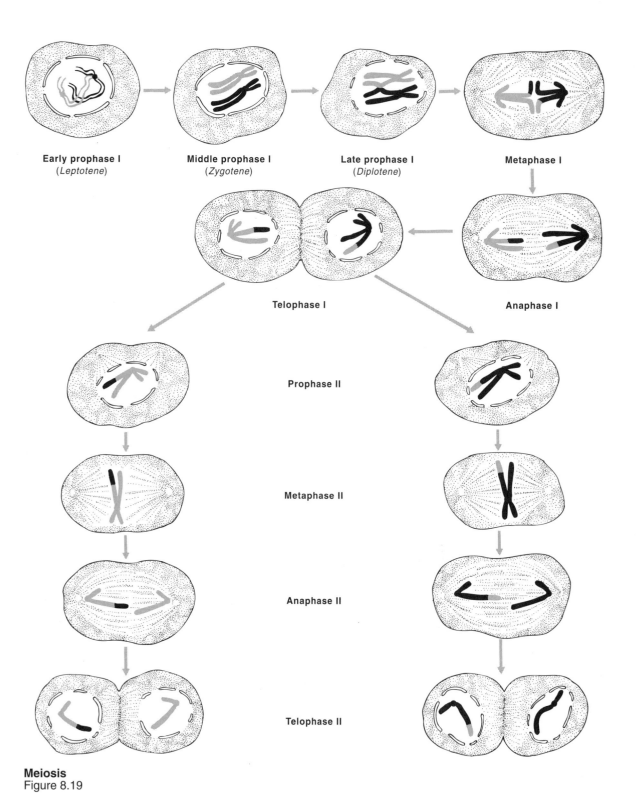

Early prophase I
(*Leptotene*)

Middle prophase I
(*Zygotene*)

Late prophase I
(*Diplotene*)

Metaphase I

Telophase I

Anaphase I

Prophase II

Metaphase II

Anaphase II

Telophase II

Meiosis
Figure 8.19

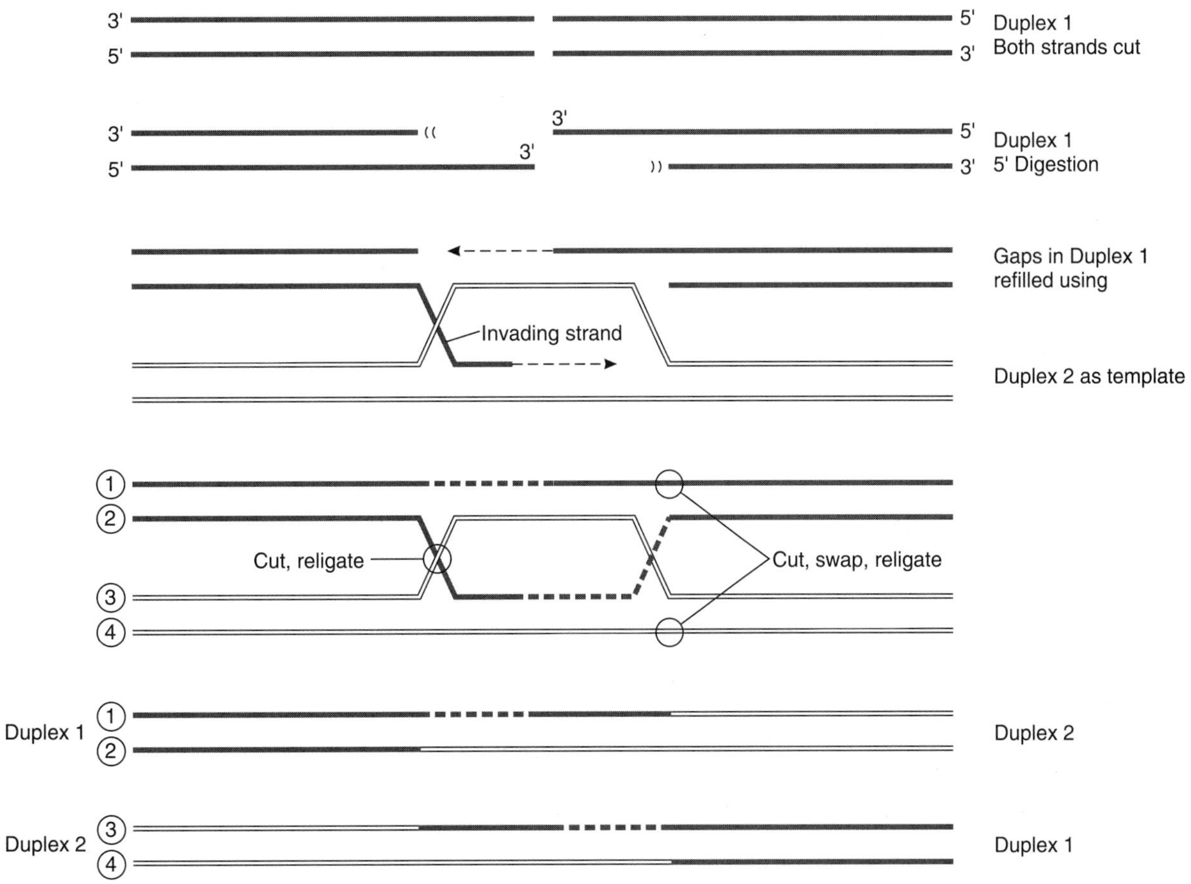
3'
5'
5'
3'
Duplex 1
Both strands cut
3'
5'
3'
3'
5'
3'
Duplex 1
5' Digestion
Invading strand
Gaps in Duplex 1
refilled using
Duplex 2 as template
1
2
3
4
Cut, religate
Cut, swap, religate
Duplex 1
1
2
Duplex 2
Duplex 2
3
4
Duplex 1

One plausible mechanism of genetic recombination
Figure 8.22

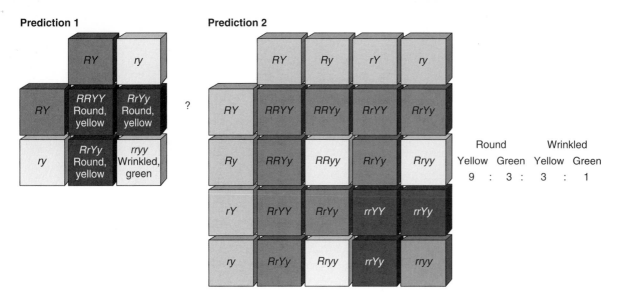

Prediction 1

	RY	ry
RY	RRYY Round, yellow	RrYy Round, yellow
ry	RrYy Round, yellow	rryy Wrinkled, green

?

Prediction 2

	RY	Ry	rY	ry
RY	RRYY	RRYy	RrYY	RrYy
Ry	RRYy	RRyy	RrYy	Rryy
rY	RrYY	RrYy	rrYY	rrYy
ry	RrYy	Rryy	rrYy	rryy

Round		Wrinkled	
Yellow	Green	Yellow	Green
9 :	3 :	3 :	1

Alternate predictions of the results of mating two dihybrids
Figure 9.7

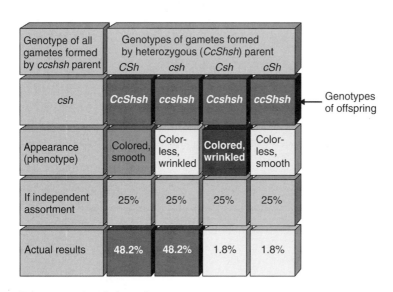

Genotype of all gametes formed by ccshsh parent	Genotypes of gametes formed by heterozygous (CcShsh) parent			
	CSh	csh	Csh	cSh
csh	CcShsh	ccshsh	Ccshsh	ccShsh
Appearance (phenotype)	Colored, smooth	Color-less, wrinkled	**Colored, wrinkled**	Color-less, smooth
If independent assortment	25%	25%	25%	25%
Actual results	**48.2%**	**48.2%**	1.8%	1.8%

← Genotypes of offspring

Demonstrating linkage in corn
Figure 9.9

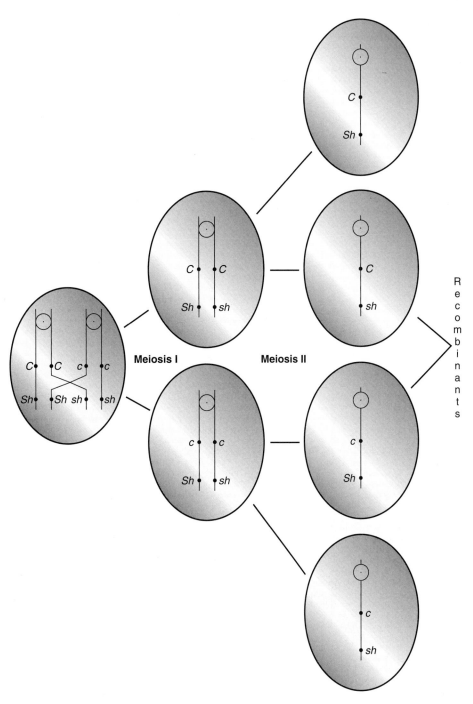

Meiosis I

Meiosis II

Recombinants

Schematic representation of the recombination of linked genes
Figure 9.10

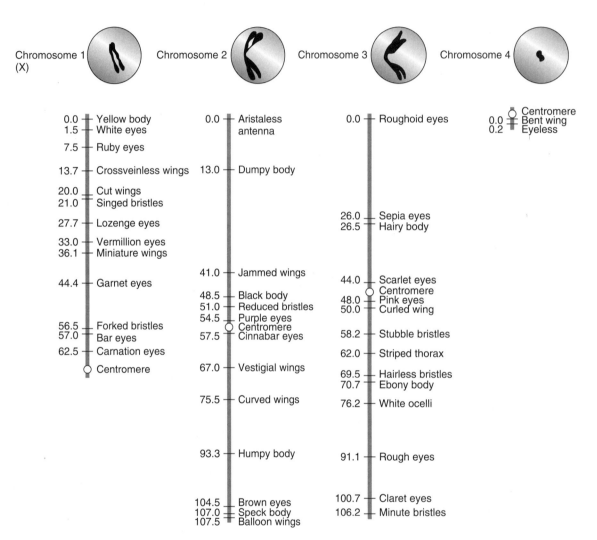

Maps of the chromosomes of *Drosophila melanogaster*
Figure 9.13

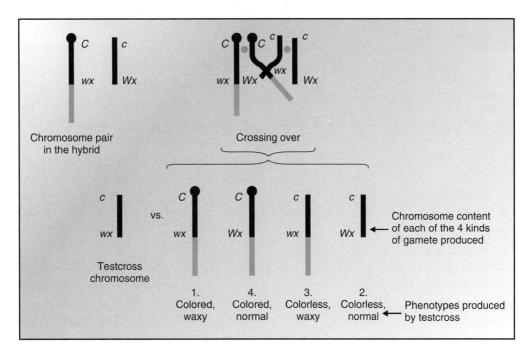

Correlation of cytological crossing-over and genetic recombination in corn
Figure 9.15

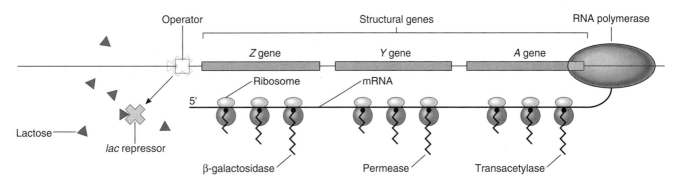

The *lac* operon
Figure 10.2

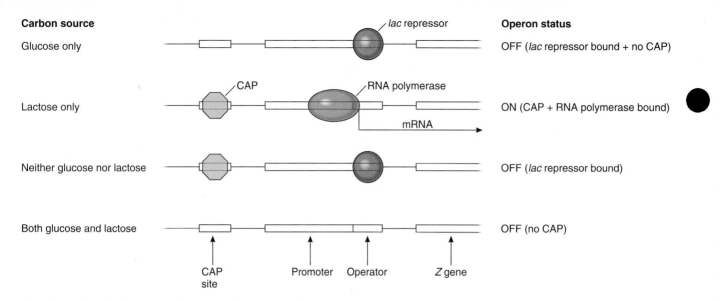

Carbon source		Operon status
Glucose only		OFF (*lac* repressor bound + no CAP)
Lactose only		ON (CAP + RNA polymerase bound)
Neither glucose nor lactose		OFF (*lac* repressor bound)
Both glucose and lactose		OFF (no CAP)

E. coli prefers glucose over lactose as its source of carbon and energy
Figure 10.4

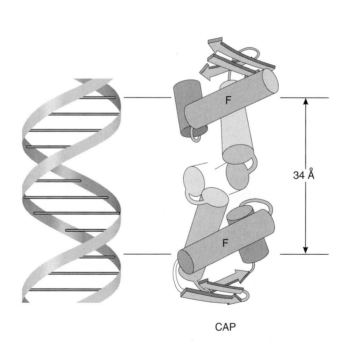

Model of the catabolite activator protein (CAP)
Figure 10.5

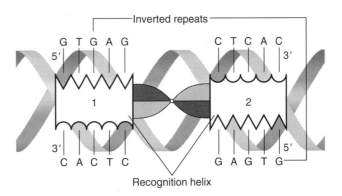

How identical recognition helices see the same sequence of base pairs in DNA
Figure 10.6

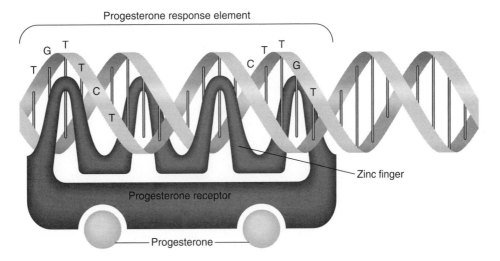

Model of the interaction between progesterone receptor and sequence of bases
Figure 10.12

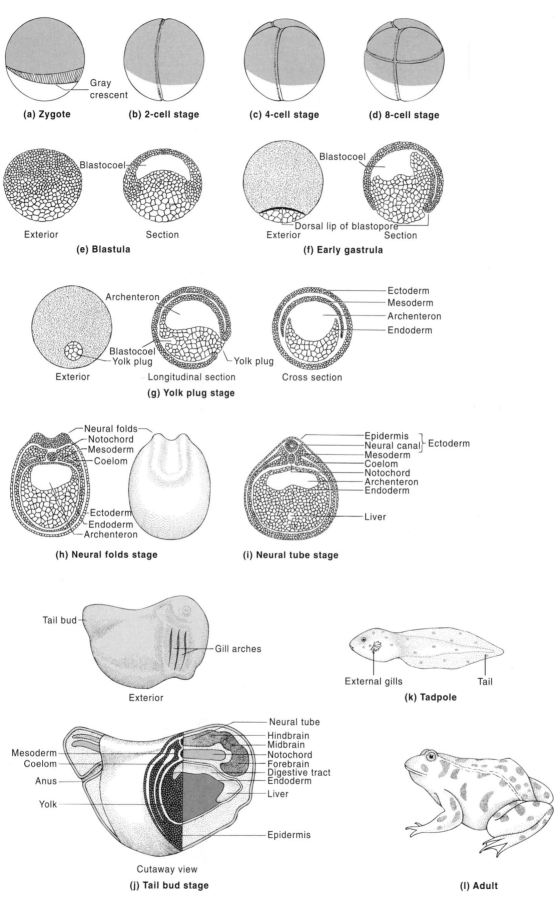

(a) Zygote

Gray crescent

(b) 2-cell stage

(c) 4-cell stage

(d) 8-cell stage

Blastocoel

Exterior   Section

(e) Blastula

Blastocoel

Dorsal lip of blastopore

Exterior   Section

(f) Early gastrula

Archenteron

Ectoderm
Mesoderm
Archenteron
Endoderm

Blastocoel
Yolk plug

Yolk plug

Exterior   Longitudinal section   Cross section

(g) Yolk plug stage

Neural folds
Notochord
Mesoderm
Coelom

Ectoderm
Endoderm
Archenteron

(h) Neural folds stage

Epidermis
Neural canal ⎫ Ectoderm
Mesoderm
Coelom
Notochord
Archenteron
Endoderm

Liver

(i) Neural tube stage

Tail bud

Gill arches

Exterior

External gills   Tail

(k) Tadpole

Neural tube
Hindbrain
Midbrain
Notochord
Forebrain
Digestive tract
Endoderm

Mesoderm
Coelom

Anus

Yolk

Liver

Epidermis

Cutaway view

(j) Tail bud stage

(l) Adult

Embryonic development of the frog
Figure 11.2

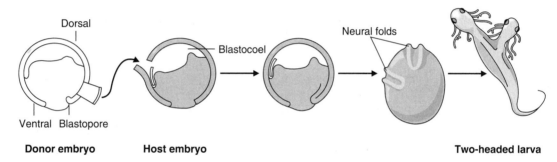

Donor embryo **Host embryo** **Two-headed larva**

The experiment of Spemann and Hilde Mangold
Figure 11.15

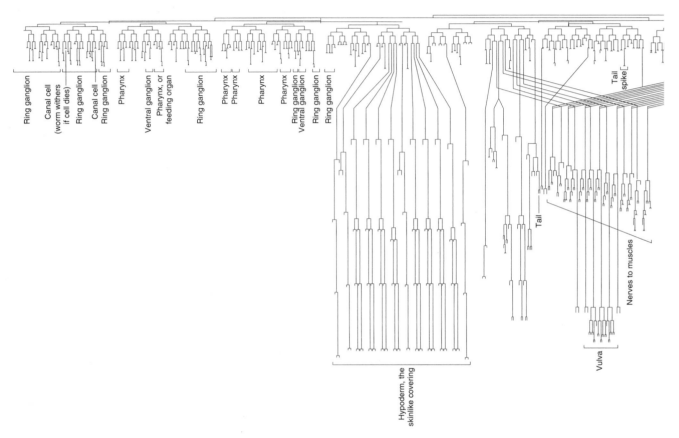

Pathways by which each of the 558 cells in the larva of
***C. elegans* has developed from the zygote**
Figure 11.20

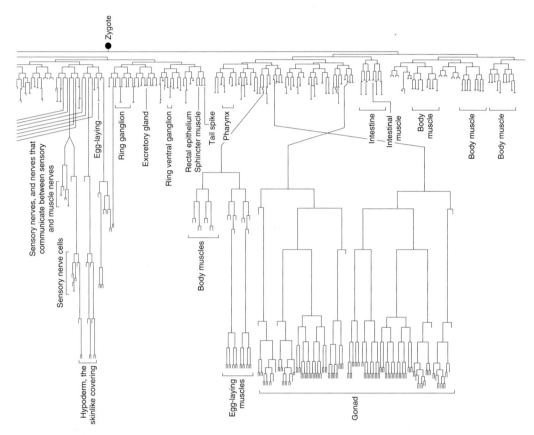

Figure 11.20 (continued)

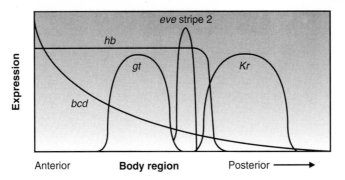

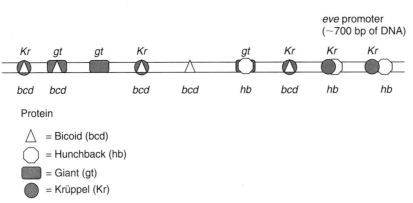

Gradients of gene expression from front to back of the developing *Drosophila* embryo
Figure 11.23

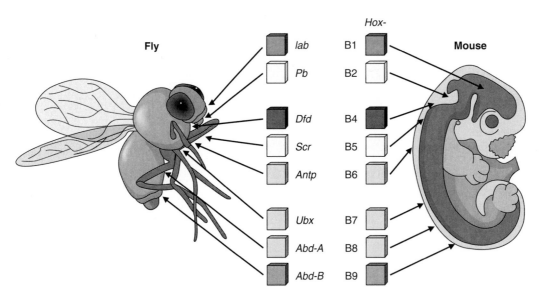

Organization of homeobox genes in *Drosophila* and the mouse
Figure 11.25

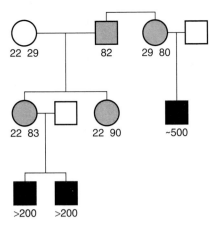

Inheritance of the fragile X syndrome in one family
Figure 12.2

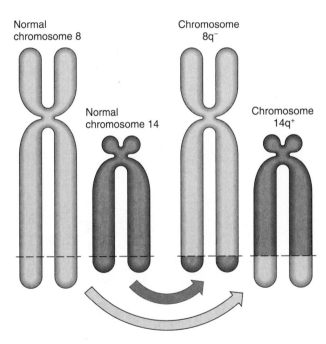

A reciprocal translocation between chromosome 8 and chromosome 14
Figure 12.5

Approximate lethal dose (in a short period to entire body)	450,000
Causes radiation sickness (when absorbed in a short period)	>100,000
Increase in lifetime dose to most heavily exposed people living near Chernobyl	43,000
Average annual dose (excluding natural background) for medical x-ray technicians	320
Maximum permissible annual dose (excluding natural background and medical exposure) to general public	170
Natural background, Boston, Mass. (per year) (excluding radon)	102
Natural background, Denver, Colo. (per year) (excluding radon)	180
Additional annual dose if you live in a brick rather than a wood house	7
Average dose to person living within 10 miles of Three-Mile Island (TMI) caused by the accident of March 28, 1979	8
Most heavily exposed person (a fisherman) near TMI	<100
Approximate dose received by a person spending 1 year at the fence surrounding a nuclear power station	0.1–0.6
Average dose to each person in the U.S. population from nuclear power plants (per year)	0.002
Received by the bone marrow during a set of dental x rays*	9.4
Annual dose to gonads from TV sets	0.2–1.5
Received by the bone marrow during a barium enema	875
Received by the bone marrow during a chest x ray	10
Received by breast during a mammogram	50–700
Average airline passenger (10 flights/year)	3
Flight crew and cabin attendants (per year)	160
Hourly dose to skin holding piece of the original "Fiesta Ware" (a brand of pottery)	200–300
Annual dose to each person in the U.S. population from fallout (former weapons testing plus Chernobyl)	0.06

*Dose much higher (several thousand mrem) to the skin in the path of the beam, but bone marrow is more susceptible to damage (e.g., leukemia).

An assortment of typical radiation doses
Figure 12.8

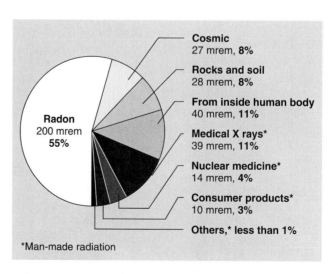

*Man-made radiation

U.S. estimated average annual radiation exposure
Figure 12.9

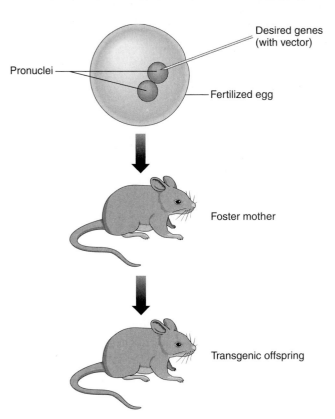

Pronuclei

Desired genes
(with vector)

Fertilized egg

Foster mother

Transgenic offspring

Making a transgenic mouse
Figure 12.13

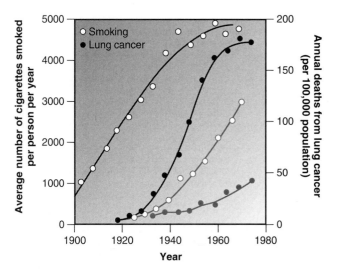

**Correlation between cigarette consumption and lung
cancer deaths among English males and females**
Figure 13.3

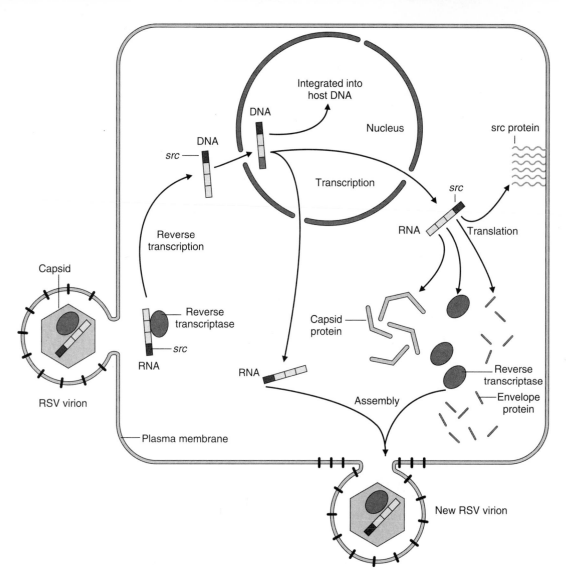

Life cycle of the Rous sarcoma virus (RSV)
Figure 13.8

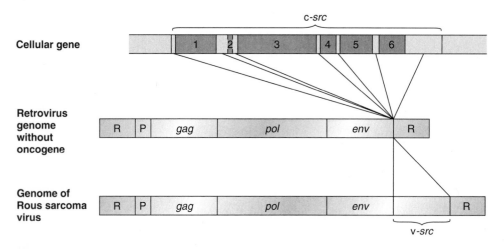

The genome of the Rous sarcoma virus
Figure 13.9

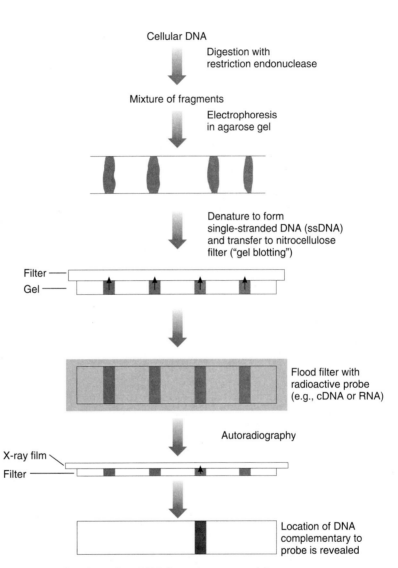

Procedure for detecting DNA fragments containing a particular sequence of nucleotides
Figure 14.1

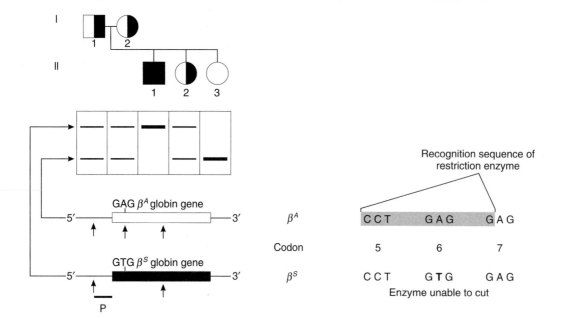

Pedigree and genotypes of a family whose only son has sickle-cell anemia
Figure 14.2

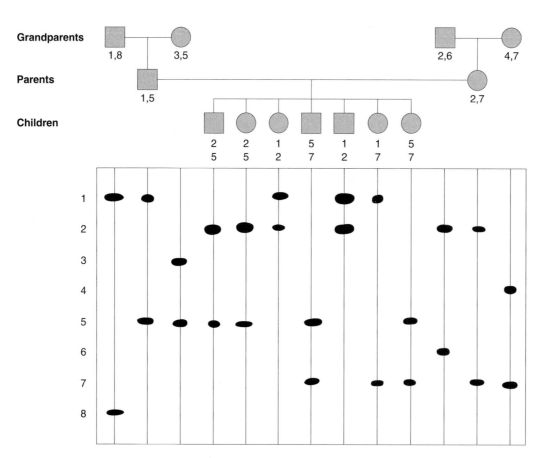

Pedigree showing the inheritance of 8 different RFLPs through three generations
Figure 14.3

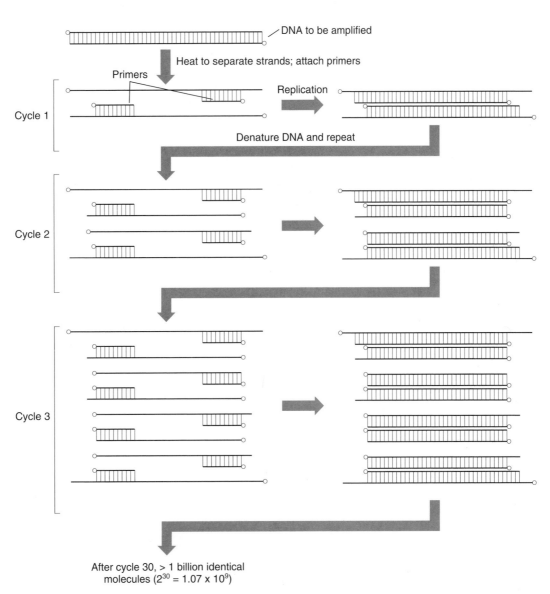

Principle of the polymerase chain reaction (PCR)
Figure 14.6

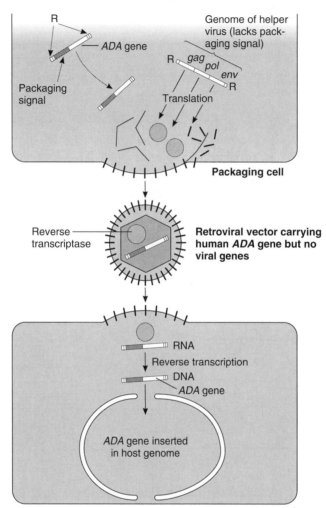

How a retrovirus can be used to introduce a therapeutic gene into the cells of the patient
Figure 14.8

Minimal Medium for *E. coli*		
glucose	5	g
Na_2HPO_4	6	g
KH_2PO_4	3	g
NH_4Cl	1	g
NaCl	0.5	g
$MgSO_4$	0.12	g
$CaCl_2$	0.01	g

(a)

Bristol's Medium for Algae		
$NaNO_3$	250	mg
K_2HPO_4	75	mg
KH_2PO_4	175	mg
$CaCl_2$	25	mg
NaCl	25	mg
$MgSO_4 \cdot 7H_2O$	75	mg
$FeCl_3$	0.3	mg
$MnSO_4 \cdot 4H_2O$	0.3	mg
$ZnSO_4 \cdot 7H_2O$	0.2	mg
H_3BO_3	0.2	mg
$CuSO_4 \cdot 5H_2O$	0.06	mg

(c)

Ham's Tissue Culture Medium for Mammalian Cells			
L-Arginine	211 mg	Biotin	0.024 mg
L-Histidine	21 mg	Calcium pantothenate	0.7 mg
L-Lysine	29.3 mg	Choline chloride	0.69 mg
L-Methionine	4.48 mg	i-Inositol	0.54 mg
L-Phenylalanine	4.96 mg	Niacinamide	0.6 mg
L-Tryptophan	0.6 mg	Pyridoxine hydrochloride	0.2 mg
L-Tyrosine	1.81 mg	Riboflavin	0.37 mg
L-Alanine	8.91 mg	Thymidine	0.7 mg
Glycine	7.51 mg	Cyanocobalamin	1.3 mg
L-Serine	10.5 mg	Sodium pyruvate	110 mg
L-Threonine	3.57 mg	Lipoic acid	0.2 mg
L-Aspartic acid	13.3 mg	$CaCl_2$	44 mg
L-Glutamic acid	14.7 mg	$MgSO_4 \cdot 7H_2O$	153 mg
L-Asparagine	15 mg	Glucose	1.1 g
L-Glutamine	146.2 mg	NaCl	7.4 g
L-Isoleucine	2.6 mg	KCl	285 mg
L-Leucine	13.1 mg	Na_2HPO_4	290 mg
L-Proline	11.5 mg	KH_2PO_4	83 mg
L-Valine	3.5 mg	Phenol red	1.2 mg
L-Cysteine	31.5 mg	$FeSO_4$	0.83 mg
Thiamine hydrochloride	1 mg	$CuSO_4 \cdot 5H_2O$	0.0025 mg
Hypoxanthine	4 mg	$ZnSO_4 \cdot 7H_2O$	0.028 mg
Folic acid	1.3 mg	$NaHCO_3$	1.2 g
		Triple distilled water	1000 ml

Recipes for media used to culture *E. coli* bacteria, human cells, and unicellular green algae
Figure 15.1

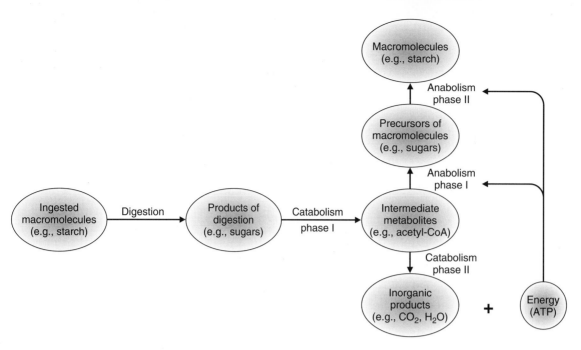

Pathways of metabolism
Figure 15.2

	♀	♂		♀	♂
Protein	44 g	56 g	Pantothenic acid	4–7 mg	Same
Vitamin A	800 µg*	1000 µg*	Biotin	100–200 µg	Same
Vitamin D	7.5 µg	Same	Calcium	800 mg	Same
Vitamin E	8 mg	10 mg	Phosphorus	800 mg	Same
Vitamin C	60 mg	Same	Magnesium	300 mg	350 mg
Thiamine	1.1 mg	1.5 mg	Iron	18 mg	10 mg
Riboflavin	1.3 mg	1.7 mg	Zinc	15 mg	Same
Niacin	14 mg	19 mg	Iodine	150 µg	Same
Vitamin B$_6$	2 µg	2.2 µg	Fluoride	1.5–4 mg	Same
Folacin	400 µg	Same	Selenium	50–200 µg	Same
Vitamin B$_{12}$	3 µg	Same			

*To the extent that the vitamin A requirement is met by ingested beta-carotene, the amount should be multiplied by 6.

**Current recommended daily dietary allowances (RDAs)
for men and women ages 19 to 22**
Figure 15.3

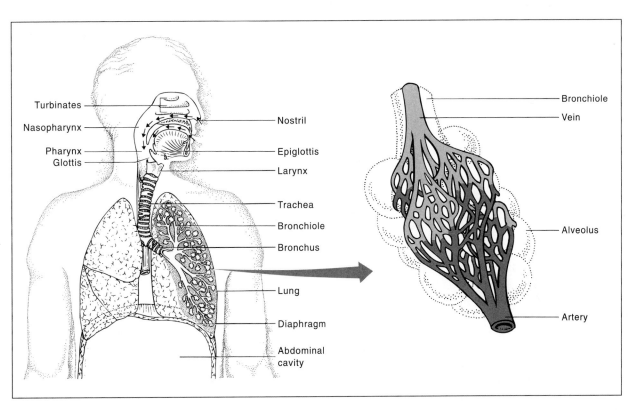

The human pulmonary system
Figure 16.6

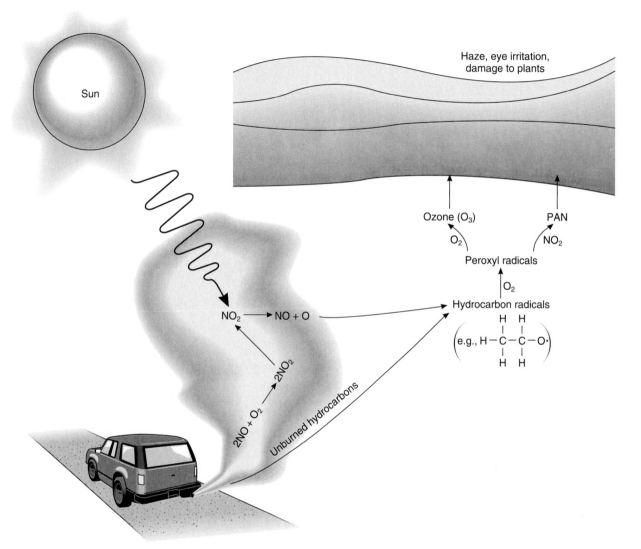

Sun

Haze, eye irritation,
damage to plants

Ozone (O_3)

PAN

O_2

NO_2

Peroxyl radicals

O_2

Hydrocarbon radicals

$$\left(\text{e.g., } H-\overset{\displaystyle H}{\underset{\displaystyle H}{C}}-\overset{\displaystyle H}{\underset{\displaystyle H}{C}}-O\cdot\right)$$

$NO_2 \longrightarrow NO + O$

$2NO_2$

$2NO + O_2$

Unburned hydrocarbons

**Representative reactions leading to the formation of
photochemical smog**
Figure 16.17

Cause of Death	Observed Deaths	Expected Deaths	Excess Deaths	Percentage of Excess	Relative Death Rate
Total deaths (all causes)	**7316**	**4651**	**2665**	**100.0**	**1.57**
Coronary artery disease	3361	1973	1388	52.1	1.70
Other heart diseases	503	425	78	2.9	1.18
Cerebral vascular lesions	556	428	128	4.8	1.30
Aneurysm and Buerger's disease	86	29	57	2.1	2.97
Other circulatory diseases	87	68	19	0.7	1.28
Lung cancer	397	37	360	13.5	10.73
Cancer of the buccal cavity, larynx or esophagus	91	18	73	2.7	5.06
Cancer of the bladder	70	35	35	1.3	2.00
Other cancers	902	651	251	9.4	1.39
Gastric and duodenal ulcer	100	25	75	2.8	4.00
Cirrhosis of the liver	83	43	40	1.5	1.93
Pulmonary disease (except cancer)	231	81	150	5.6	2.85
All other diseases	486	453	33	1.2	1.07
Accident, violence, suicide	363	385	−22	−0.8	0.94

Number of deaths of cigarette smokers compared with the number to be expected among nonsmokers of the same ages
Figure 16.21

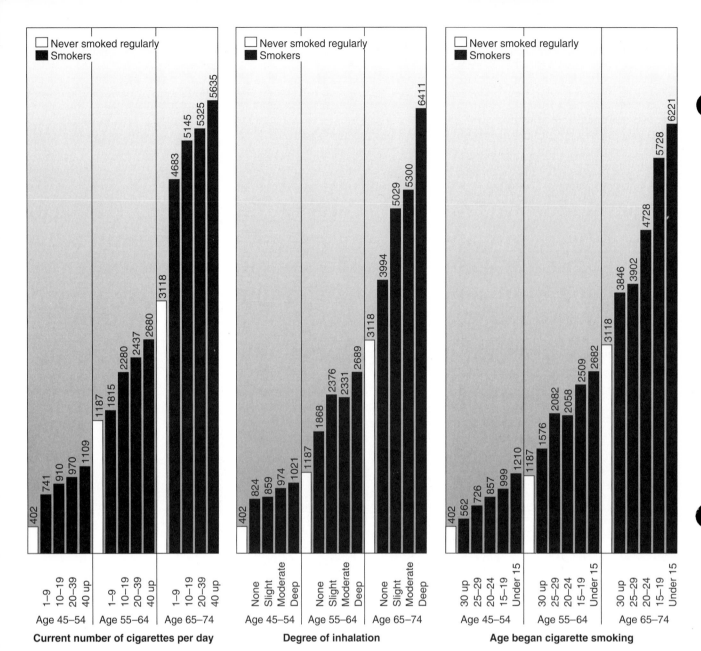

Death rates from all causes in three 10-year age brackets of nonsmokers compared with cigarette smokers
Figure 16.25

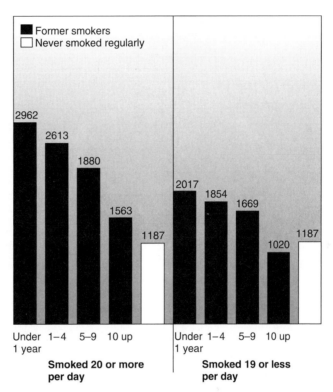

Death rates of male smokers ages 55–64 who stopped smoking for the number of years indicated
Figure 16.26

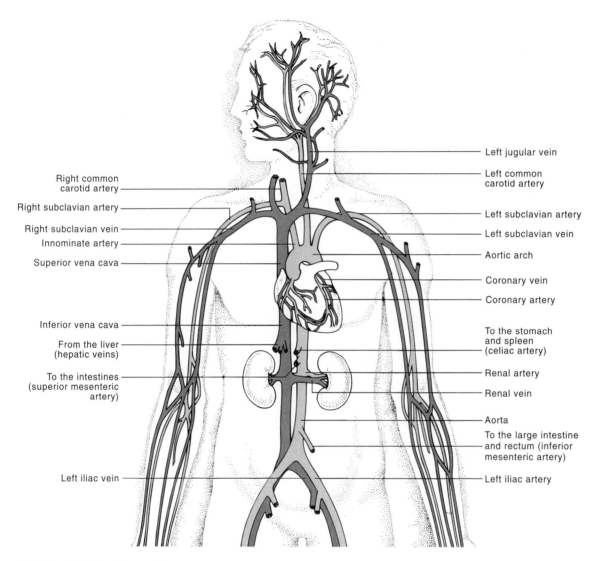

Major human blood vessels
Figure 17.4

Left jugular vein

Left common
carotid artery

Right common
carotid artery

Left subclavian artery

Right subclavian artery

Left subclavian vein

Right subclavian vein

Innominate artery

Aortic arch

Superior vena cava

Coronary vein

Coronary artery

Inferior vena cava

To the stomach
and spleen
(celiac artery)

From the liver
(hepatic veins)

Renal artery

To the intestines
(superior mesenteric
artery)

Renal vein

Aorta

To the large intestine
and rectum (inferior
mesenteric artery)

Left iliac vein

Left iliac artery

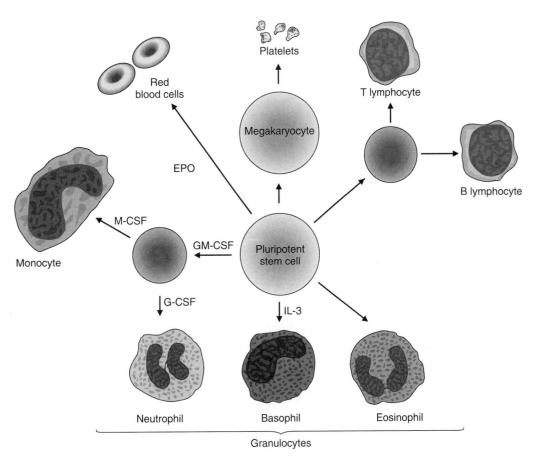

Pathways of differentiation of blood cells and their derivatives
Figure 17.10

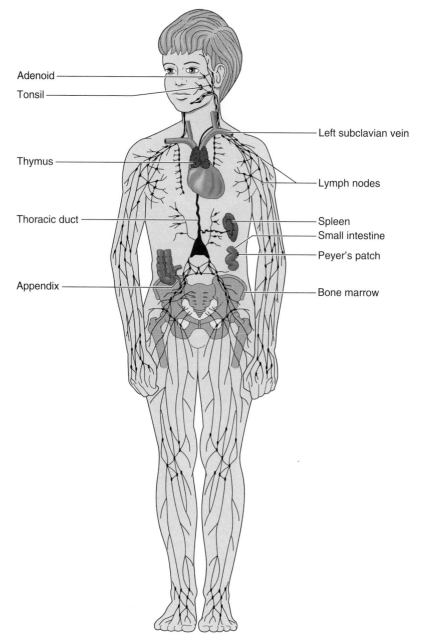

Adenoid
Tonsil
Left subclavian vein
Thymus
Lymph nodes
Thoracic duct
Spleen
Small intestine
Peyer's patch
Appendix
Bone marrow

The immune system
Figure 18.1

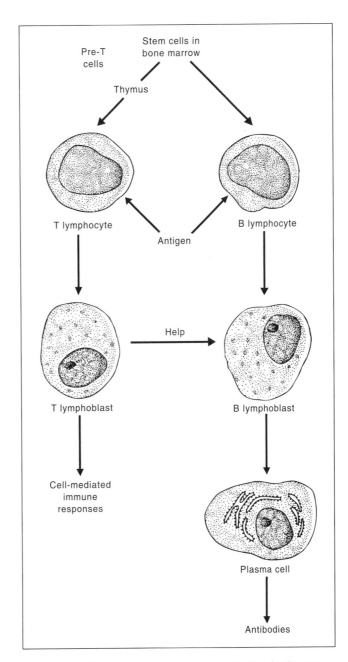

Pathways of lymphocyte development active in the two branches of the immune system
Figure 18.2

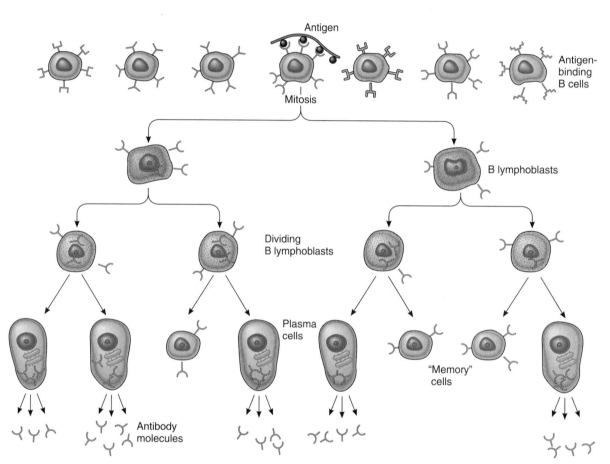

Activation of a B cell that has receptors specific for an epitope on the antigen
Figure 18.4

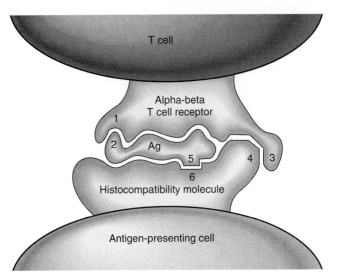

Model of the interaction between a T cell receptor for antigen, the antigen (Ag), and a histocompatibility molecule
Figure 18.7

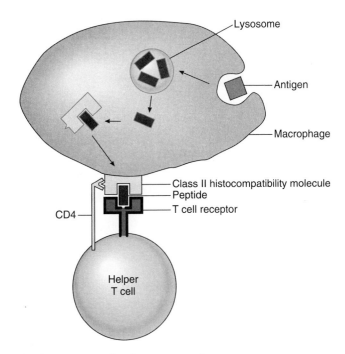

Antigen processing by a macrophage
Figure 18.10

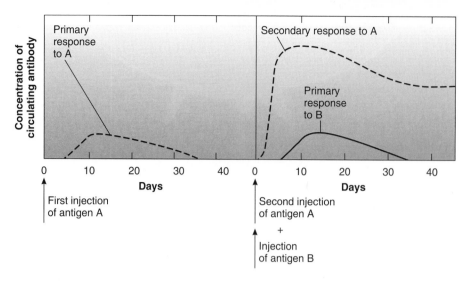

Primary and secondary antibody responses
Figure 18.16

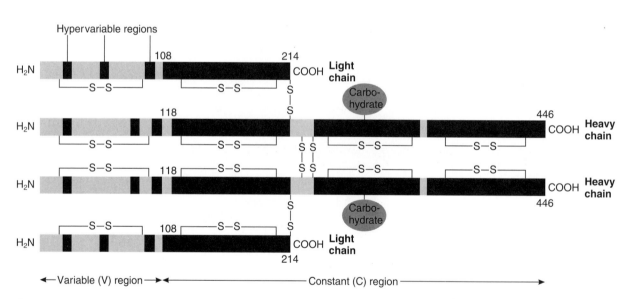

Polypeptide chain structure of an antibody molecule
Figure 18.21

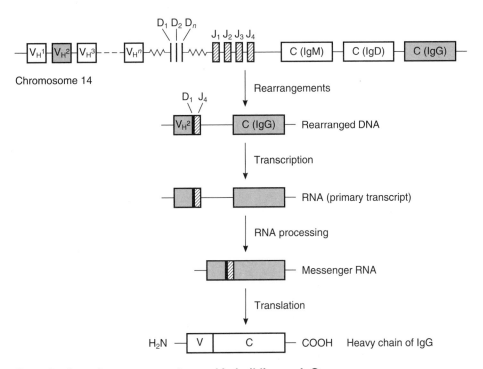

Organization of gene segments used in building an IgG molecule
Figure 18.25

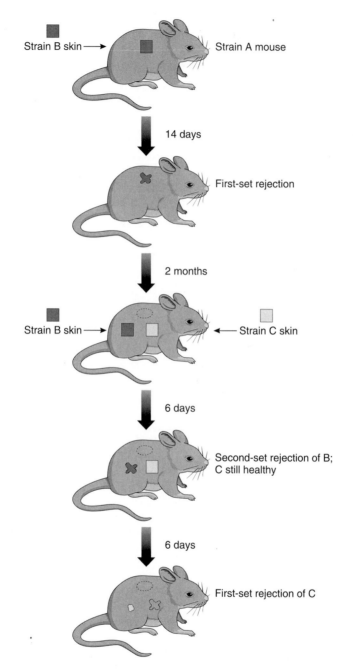

Strain B skin → | Strain A mouse

↓ 14 days

First-set rejection

↓ 2 months

Strain B skin → | ← Strain C skin

↓ 6 days

Second-set rejection of B;
C still healthy

↓ 6 days

First-set rejection of C

Demonstration that graft rejection is an immune response
Figure 18.34

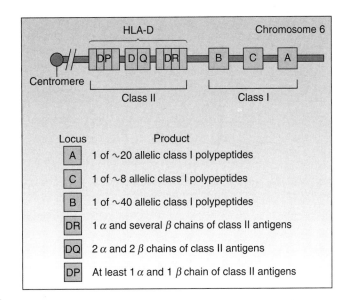

Some of the loci found in the HLA region of humans and their gene products
Figure 18.36

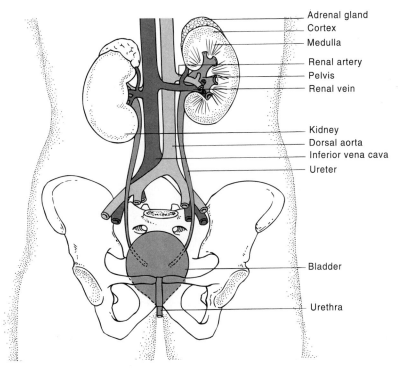

The human excretory system
Figure 19.1

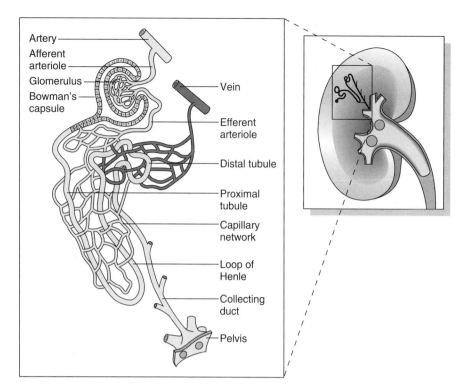

A single nephron
Figure 19.2

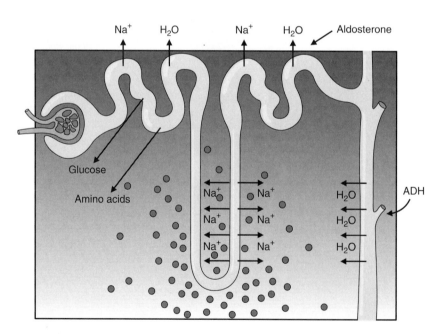

Transport activities in the nephron
Figure 19.5

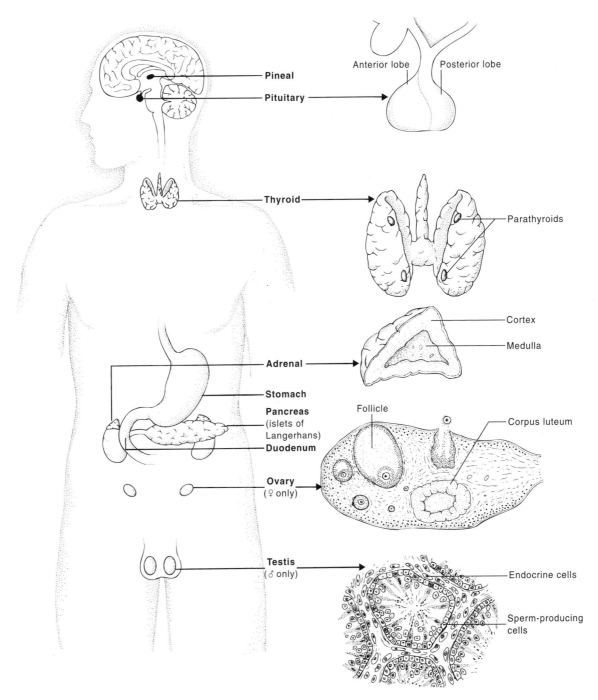

Human endocrine glands
Figure 19.8

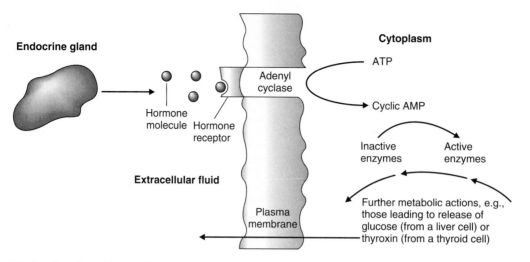

Mechanism by which polypeptide hormones stimulate their target cells
Figure 19.13

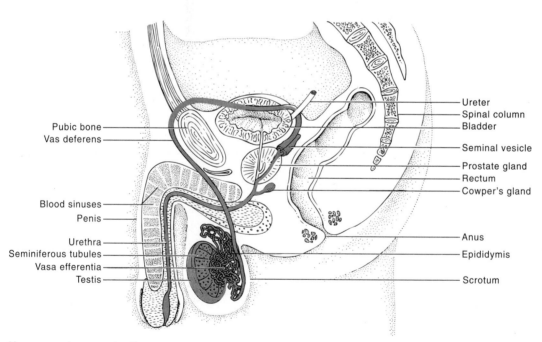

Human male reproductive organs
Figure 20.1

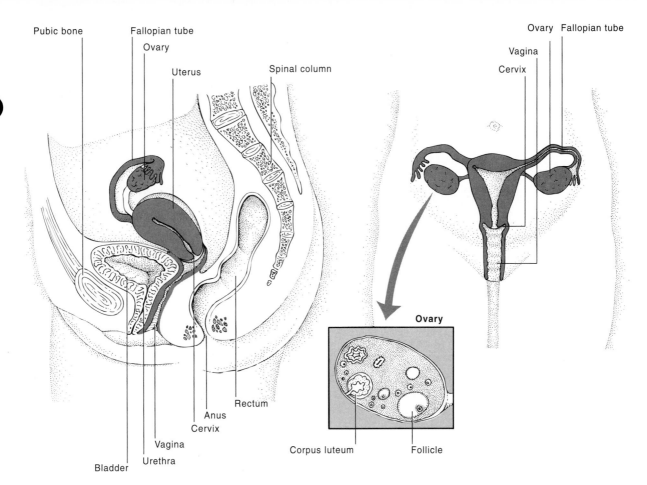

Human female reproductive organs
Figure 20.4

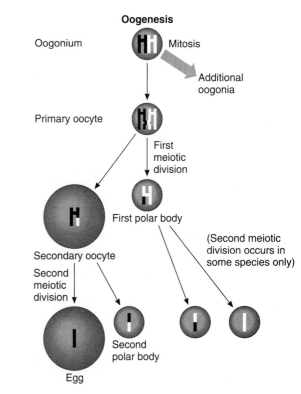

Egg formation (oogenesis)
Figure 20.5

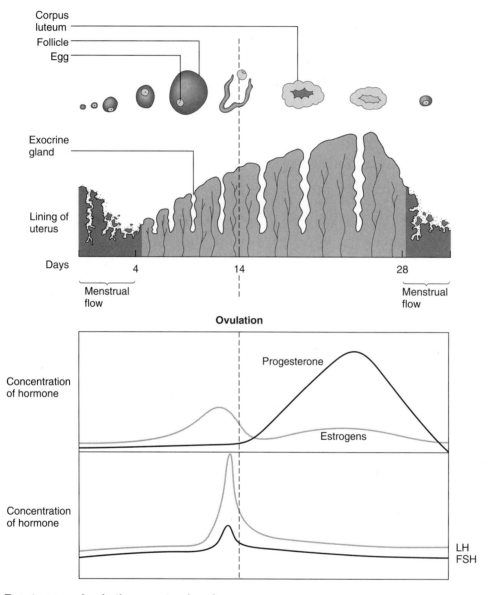

Corpus luteum

Follicle

Egg

Exocrine gland

Lining of uterus

Days

4   14   28

Menstrual flow   **Menstrual flow**

Ovulation

Concentration of hormone

Progesterone

Estrogens

Concentration of hormone

LH
FSH

Events occurring in the menstrual cycle
Figure 20.6

74

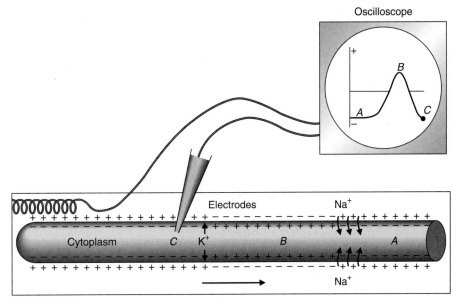

Oscilloscope

Electrodes

Na⁺

Cytoplasm

Na⁺

The nerve impulse (action potential)
Figure 21.5

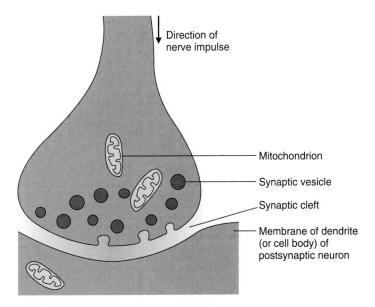

Direction of
nerve impulse

Mitochondrion

Synaptic vesicle

Synaptic cleft

Membrane of dendrite
(or cell body) of
postsynaptic neuron

Generalized structure of a synapse
Figure 21.6

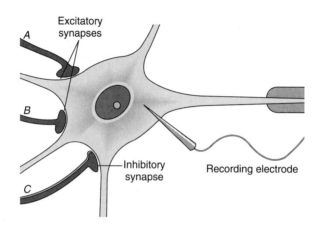

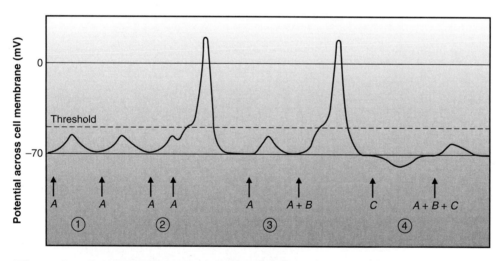

Effects of excitatory postsynaptic potentials (EPSPs) and inhibitory postsynaptic potentials (IPSPs) on the creation of action potentials in a motor neuron
Figure 21.8

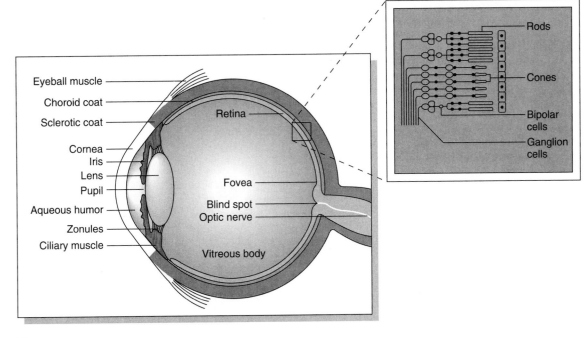

Enlarged portion of retina (schematic)

Rods
Cones
Bipolar cells
Ganglion cells

Eyeball muscle
Choroid coat
Sclerotic coat
Cornea
Iris
Lens
Pupil
Aqueous humor
Zonules
Ciliary muscle
Retina
Fovea
Blind spot
Optic nerve
Vitreous body

The human eye
Figure 21.19

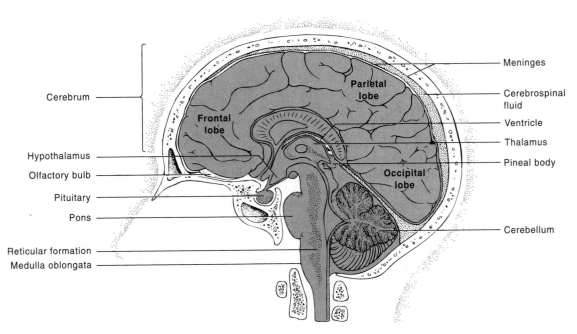

Cerebrum
Hypothalamus
Olfactory bulb
Pituitary
Pons
Reticular formation
Medulla oblongata
Parietal lobe
Frontal lobe
Occipital lobe
Meninges
Cerebrospinal fluid
Ventricle
Thalamus
Pineal body
Cerebellum

The human brain cut lengthwise between the two cerebral hemispheres
Figure 22.2

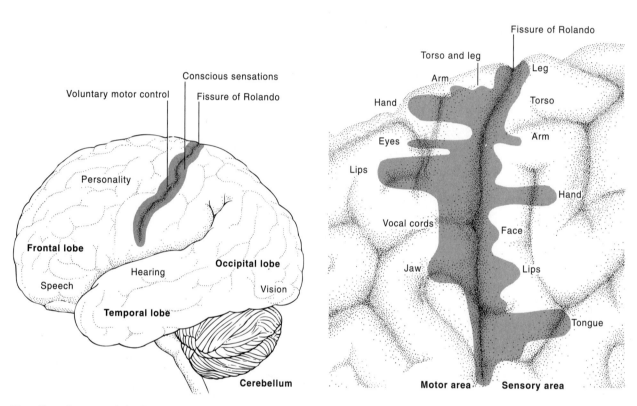

Functional areas of the human cerebrum
Figure 22.3

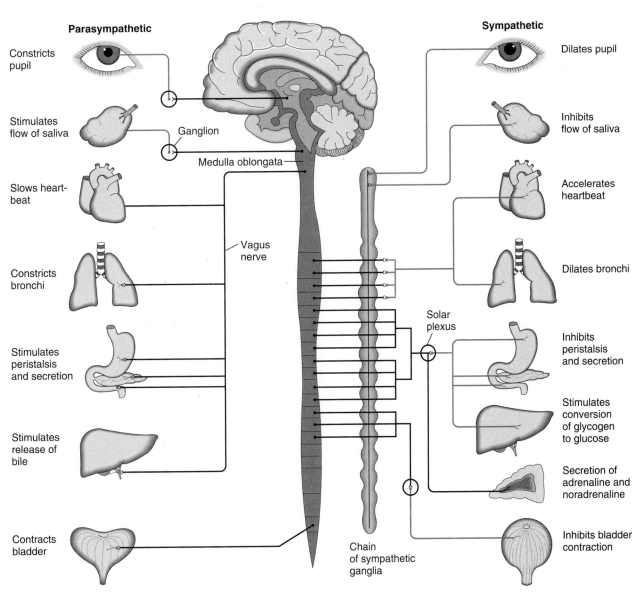

Parasympathetic

Constricts pupil

Stimulates flow of saliva

Ganglion

Slows heart-beat

Medulla oblongata

Vagus nerve

Constricts bronchi

Stimulates peristalsis and secretion

Stimulates release of bile

Contracts bladder

Sympathetic

Dilates pupil

Inhibits flow of saliva

Accelerates heartbeat

Dilates bronchi

Solar plexus

Inhibits peristalsis and secretion

Stimulates conversion of glycogen to glucose

Secretion of adrenaline and noradrenaline

Inhibits bladder contraction

Chain of sympathetic ganglia

The autonomic nervous system
Figure 22.12

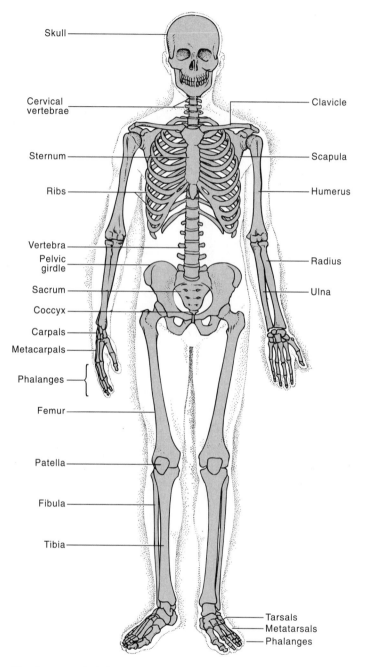

Skull

Cervical
vertebrae

Sternum

Ribs

Vertebra

Pelvic
girdle

Sacrum

Coccyx

Carpals

Metacarpals

Phalanges

Femur

Patella

Fibula

Tibia

Clavicle

Scapula

Humerus

Radius

Ulna

Tarsals
Metatarsals
Phalanges

The human skeleton
Figure 23.1

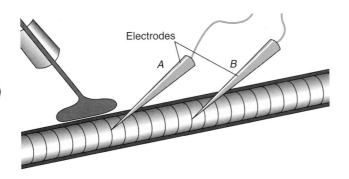

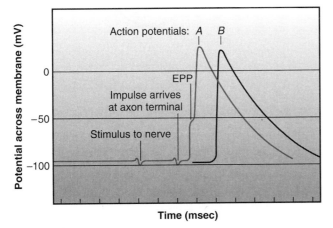

Intracellular recording from a muscle fiber being stimulated by its motor neuron
Figure 23.6

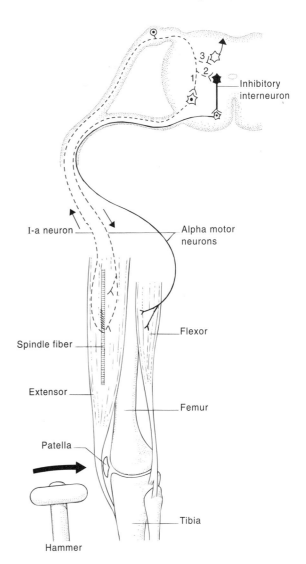

Mechanism of the knee jerk reflex, a stretch reflex
Figure 23.17

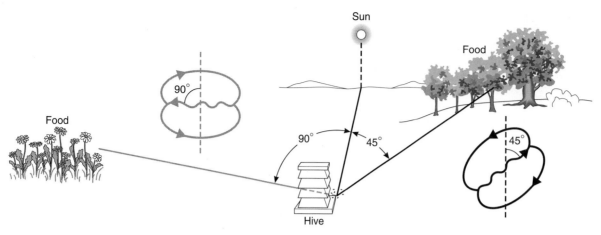

How a honeybee communicates the direction of food
Figure 24.17

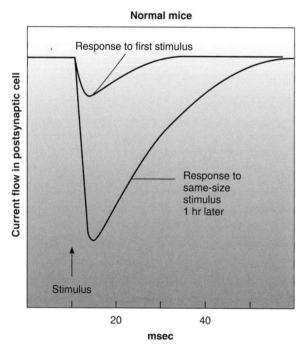

Normal mice

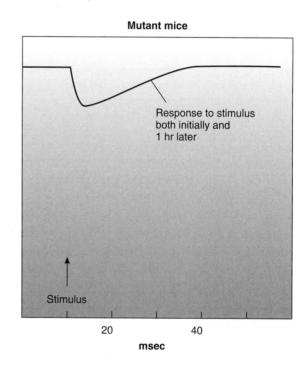

Mutant mice

Measuring long-term potentiation
Figure 24.32

82

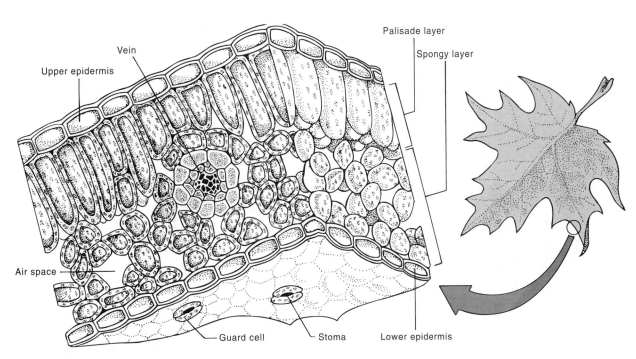

Structure of a typical leaf as seen in cross section
Figure 25.1

Upper epidermis
Vein
Palisade layer
Spongy layer
Air space
Guard cell
Stoma
Lower epidermis

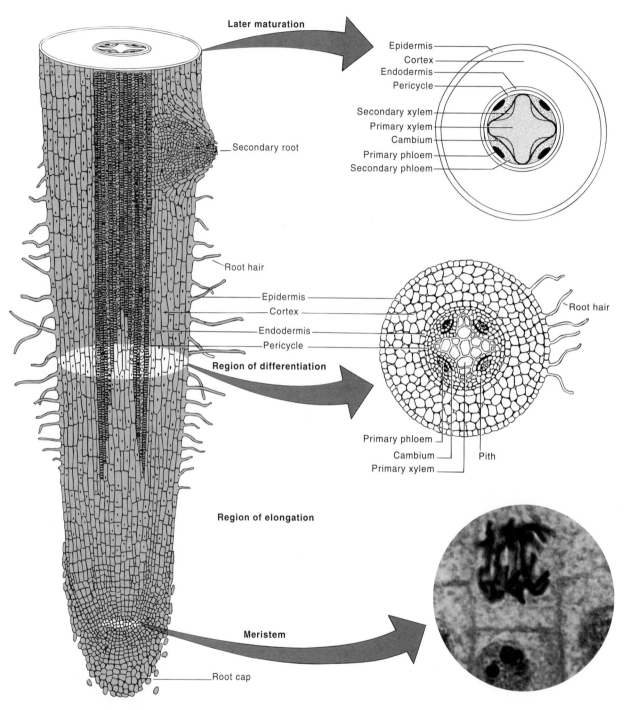

Later maturation

Epidermis
Cortex
Endodermis
Pericycle
Secondary xylem
Primary xylem
Cambium
Primary phloem
Secondary phloem

Secondary root

Root hair

Epidermis
Cortex
Endodermis
Pericycle

Region of differentiation

Root hair

Primary phloem
Cambium
Primary xylem
Pith

Region of elongation

Meristem

Root cap

**Organization of a young dicot root in longitudinal
section and cross sections**
Figure 25.8

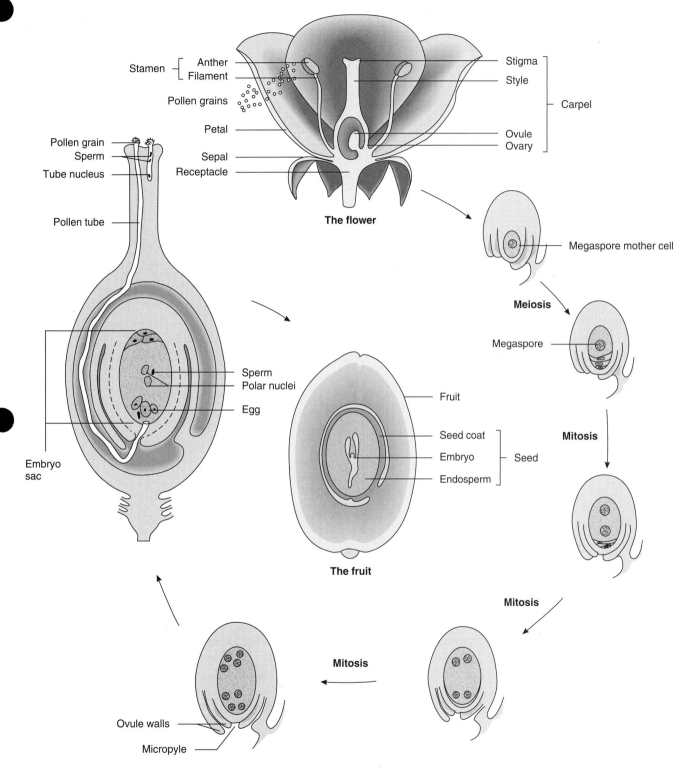

Stamen {
Anther
Filament

Pollen grains

Petal

Sepal

Receptacle

The flower

Stigma
Style
} Carpel
Ovule
Ovary

Pollen grain
Sperm
Tube nucleus

Pollen tube

Sperm
Polar nuclei

Egg

Embryo
sac

Megaspore mother cell

Meiosis

Megaspore

Mitosis

Fruit

Seed coat
Embryo   } Seed
Endosperm

The fruit

Mitosis

Mitosis

Ovule walls

Micropyle

Generalized life cycle of an angiosperm
Figure 26.3

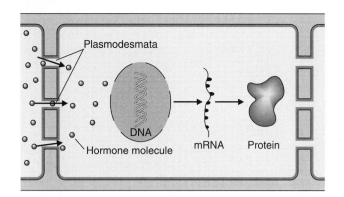

The arrival of a plant hormone triggers gene expression
in the cell
Figure 27.19

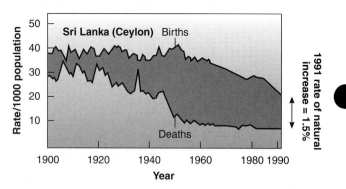

Birth and death rates in Ceylon since 1900
Figure 28.5

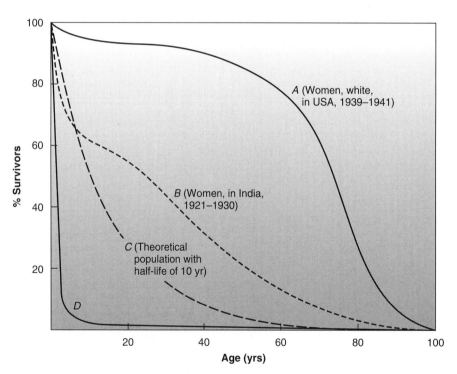

Representative survival curves
Figure 28.18

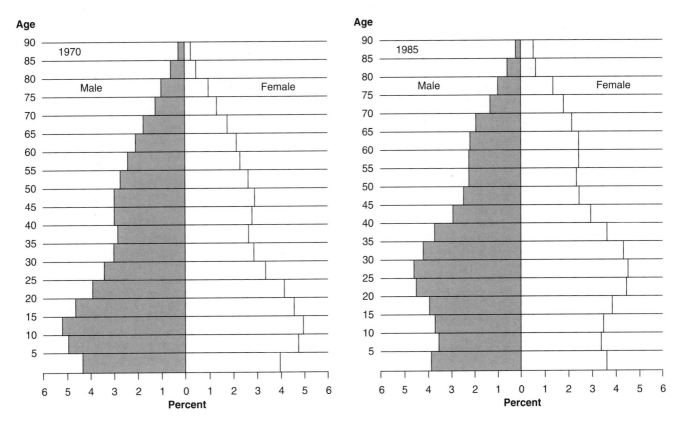

Pyramids of the U.S. population in 1970 and 1985
Figure 28.21

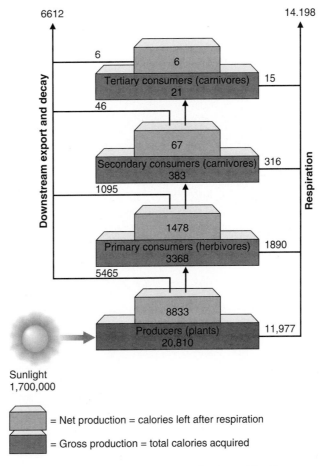

The flow of energy through Silver Springs, Florida
Figure 29.10

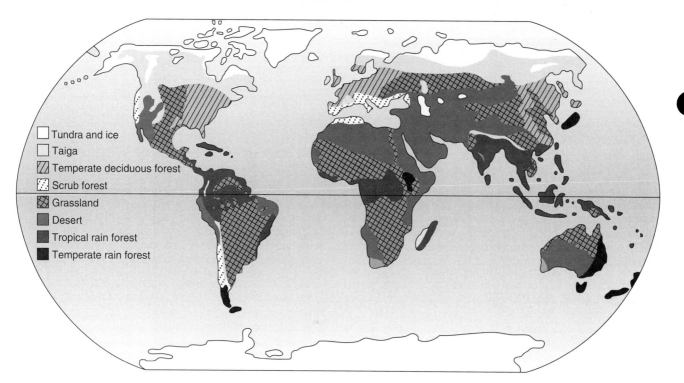

The major biomes of the world
Figure 29.14

Legend:
- Tundra and ice
- Taiga
- Temperate deciduous forest
- Scrub forest
- Grassland
- Desert
- Tropical rain forest
- Temperate rain forest

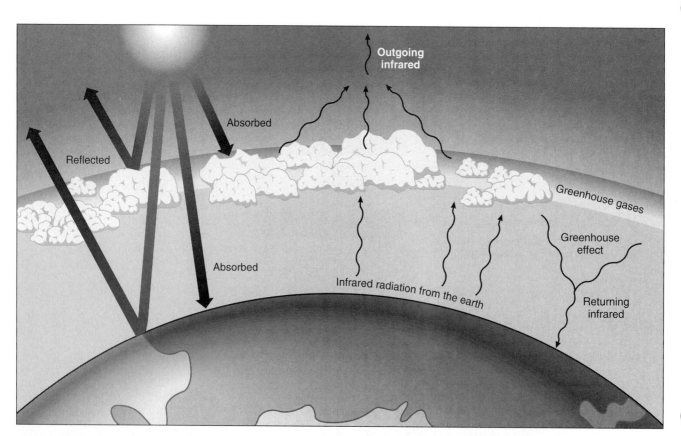

Mechanism of the greenhouse effect
Figure 30.4

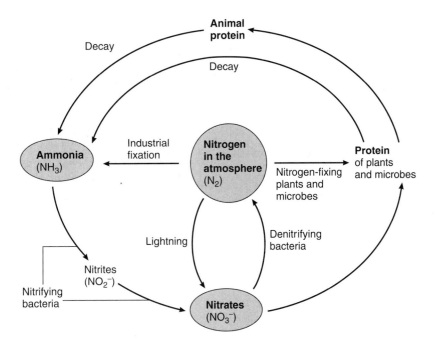

The nitrogen cycle
Figure 30.8

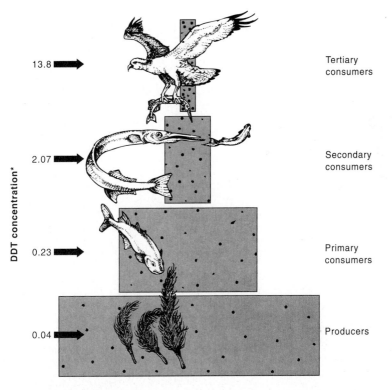

* Representative values of the concentration in the tissues of DDT and its derivatives (in parts per million, ppm).

Concentration of DDT in four successive trophic levels of a food chain
Figure 31.4

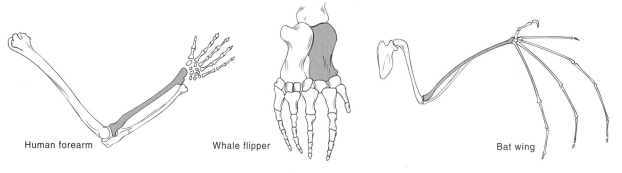

Human forearm   Whale flipper   Bat wing

Three vertebrate forelimbs: a study in homology
Figure 32.5

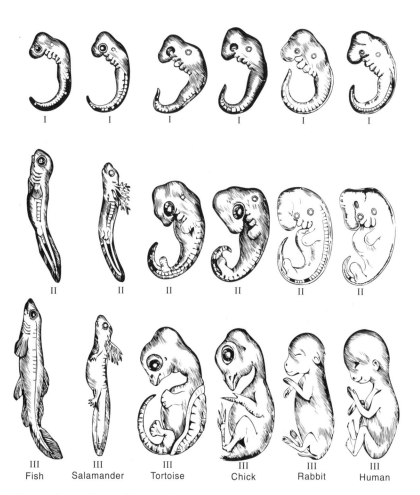

I   I   I   I   I   I

II   II   II   II   II   II

III   III   III   III   III   III
Fish   Salamander   Tortoise   Chick   Rabbit   Human

Comparison of vertebrate embryos
Figure 32.7

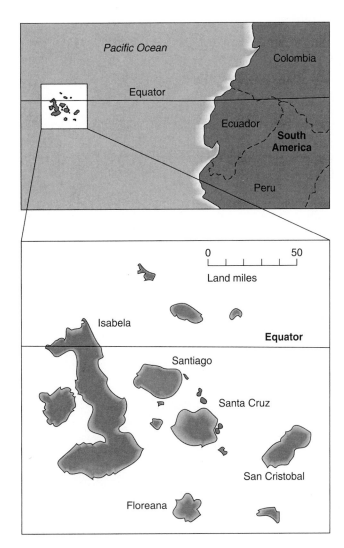

The Galápagos Islands
Figure 32.19

Darwin's finches
Figure 32.20

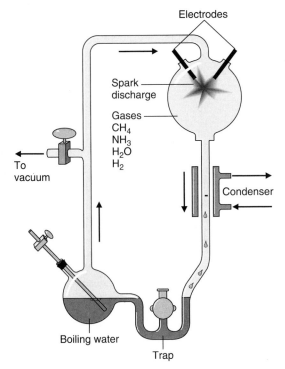

Miller's apparatus
Figure 33.3

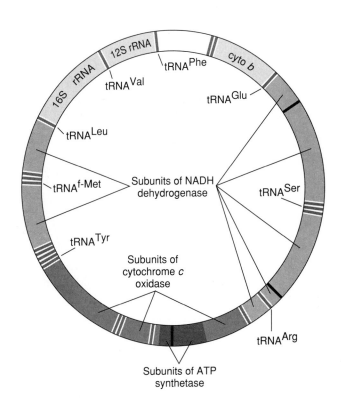

The genes of human mitochondrial DNA
Figure 33.8

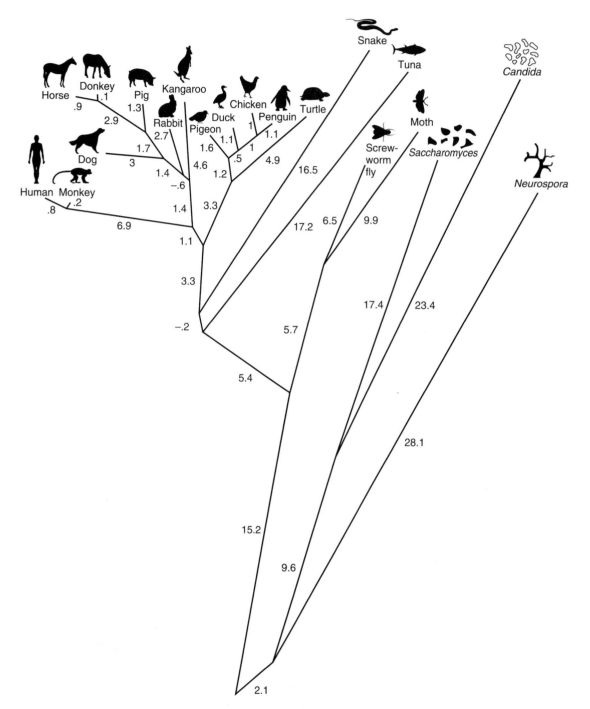

**Evolutionary tree based on the amino acid sequences
of cytochrome _c_ in 20 species**
Figure 33.17

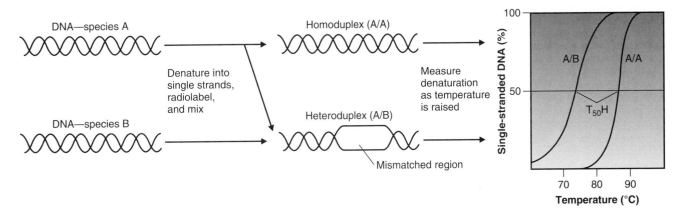

Molecular basis of DNA/DNA hybridization studies
Figure 33.18

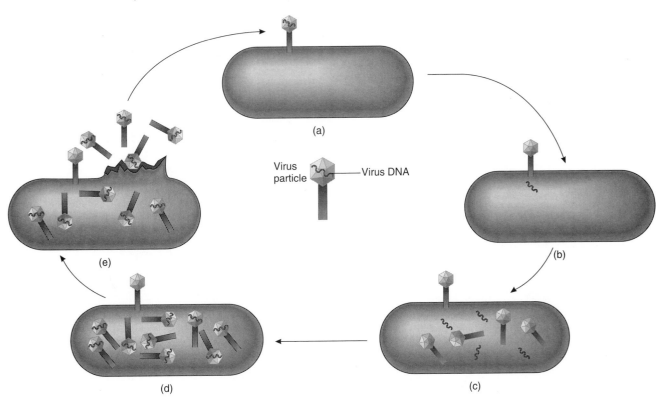

Cycle of a lytic infection of a bacterial cell by a DNA bacteriophage
Figure 34.11

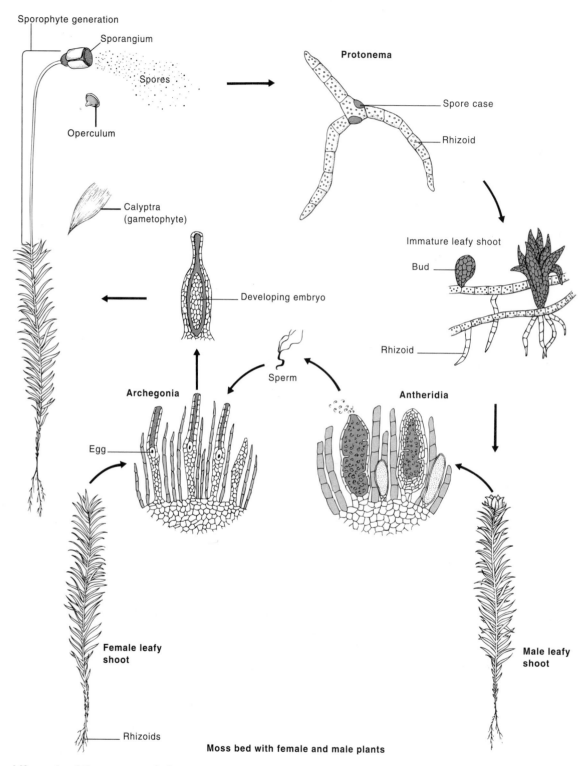

Sporophyte generation

Sporangium

Spores

Operculum

Protonema

Spore case

Rhizoid

Calyptra
(gametophyte)

Immature leafy shoot

Bud

Rhizoid

Developing embryo

Sperm

Archegonia

Antheridia

Egg

**Female leafy
shoot**

**Male leafy
shoot**

Rhizoids

Moss bed with female and male plants

**Life cycle of the common haircap moss *Polytrichum
commune***
Figure 36.4

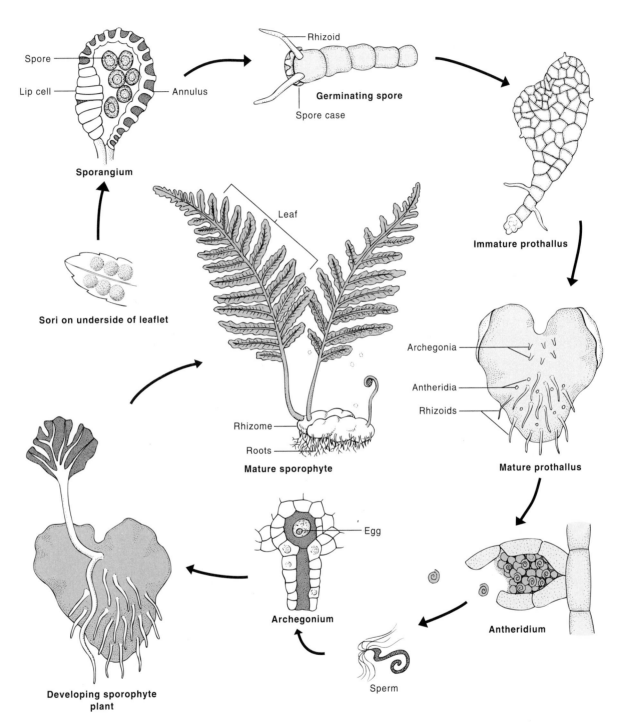

Spore

Lip cell

Annulus

Sporangium

Rhizoid

Germinating spore

Spore case

Immature prothallus

Leaf

Sori on underside of leaflet

Archegonia

Antheridia

Rhizoids

Rhizome

Roots

Mature sporophyte

Mature prothallus

Egg

Archegonium

Antheridium

Sperm

Developing sporophyte plant

Life cycle of a typical fern
Figure 36.10

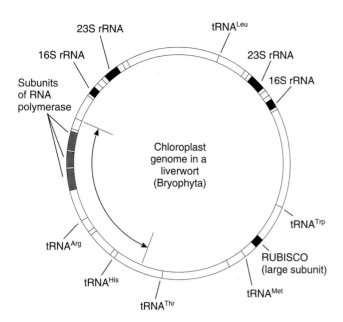

Some of the 128 genes encoded by the DNA of the chloroplasts in *Marchantia polymorpha*
Figure 36.18

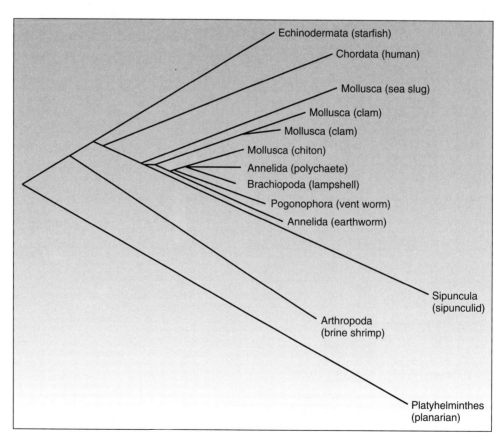

One evolutionary tree generated by computer analysis of the 18S rRNA of representatives from several animal phyla
Figure 37.24

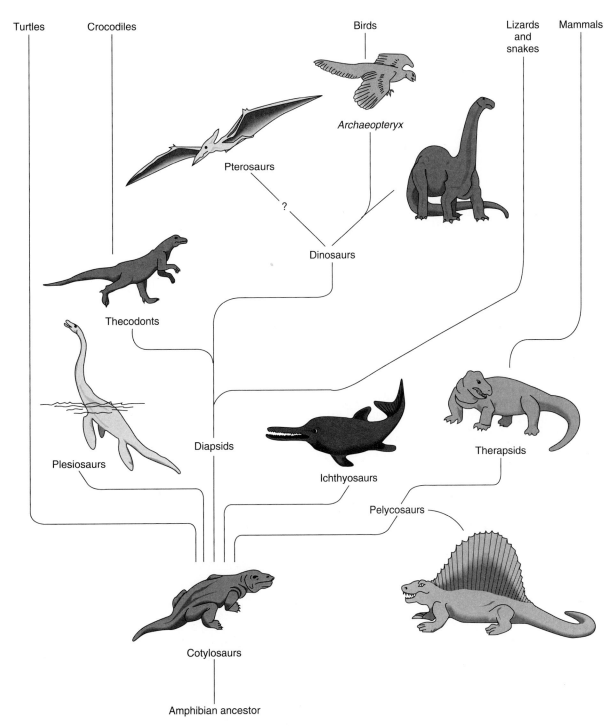

A reconstruction of reptile evolution
Figure 38.7

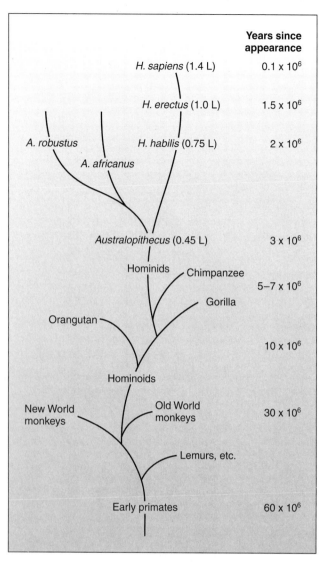

Years since appearance

H. sapiens (1.4 L)	0.1 x 10⁶
H. erectus (1.0 L)	1.5 x 10⁶
A. robustus *H. habilis* (0.75 L)	2 x 10⁶
A. africanus	
Australopithecus (0.45 L)	3 x 10⁶
Hominids Chimpanzee	5–7 x 10⁶
Gorilla	
Orangutan	10 x 10⁶
Hominoids	
New World monkeys Old World monkeys	30 x 10⁶
Lemurs, etc.	
Early primates	60 x 10⁶

Possible evolutionary relationships of the primates
Figure 38.17

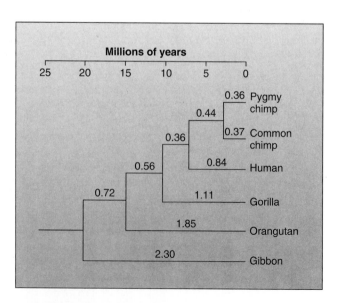

A phylogenetic tree of living hominoids based on DNA/DNA hybridization data
Figure 38.20

CREDITS

Line Art

Fig. 6.11: From William Etkin, "A Representative of the Structure of DNA" in *Bioscience,* 23:653, 1973. Copyright © 1973 American Institute of Biological Sciences, Washington, DC. Reprinted by permission.

Fig. 6.13: From Thomas D. Gelehrter, *Principles of Medical Genetics.* Copyright © 1990 Williams & Wilkins, Baltimore, MD. Reprinted by permission.

Fig. 6.26: From David A. Miklos and Greg A. Freyer, *DNA Science: A First Course in Recombinant DNA Technology.* Copyright © 1990 Cold Spring Harbor Laboratory Press, Cold Spring Harbor, New York and Carolina Biological Supply Company, Burlington, North Carolina. Reprinted by permission.

Fig. 8.14: From Robert F. Weaver and Philip W. Hedrick, *Genetics,* 2d ed. Copyright © 1992 Wm. C. Brown Communications, Inc., Dubuque, Iowa. All Rights Reserved. Reprinted by permission.

Fig. 11.20: From J. Sulston, et al., *Monograph* 17. Copyright © Cold Spring Harbor Laboratory, Cold Spring Harbor, NY. Reprinted by permission.

Fig. 11.23 (top): From D. Stanojevic, et al., "Regulation of a Segmentation Stripe by Overlapping Activators and Repressors in the *Drosophila* Embryo" in *Science,* 254:1385, 1991. Copyright © 1991 American Association for the Advancement of Science, Washington, DC. Reprinted by permission.

Fig. 11.23 (bottom): From P. A. Lawrence, *The Making of a Fly.* Copyright © 1992 Blackwell Scientific Publications Ltd., Oxford, England. Reprinted by permission.

Fig. 12.2: From C. T. Caskey, et al., "Triplet Repeat Mutations in Human Disease" in *Science,* 256:784, 1992. Copyright © 1992 American Association for the Advancement of Science, Washington, DC. Reprinted by permission.

Fig. 12.9: Data from National Council on Radiation Protection and Measurements, Bethesda, MD.

Fig. 14.1: Reprinted with the permission of Macmillan College Publishing Company from *Introduction to Immunology,* Third Edition by John W. Kimball. Copyright © 1990 by Macmillan College Publishing Company, Inc.

Fig. 14.2: From S. E. Antonarakis, "Diagnosis of Genetic Disorders at the DNA Level" in *New England Journal of Medicine,* 320:153, Jan 19, 1989. Copyright © 1989 Massachusetts Medical Society, Waltham, MA. Reprinted with permission.

Fig. 15.3: Adapted with permission from *Recommended Dietary Allowances: 10th Edition.* Copyright 1989 by the National Academy of Sciences. Courtesy of the National Academy Press, Washington, D.C.

Fig. 16.21: Data from E. C. Hammond and D. Horn, 1966, appearing in John W. Kimball, *Man and Nature.* Copyright © 1975 Addison Wesley, Menlo Park, CA.

Fig. 16.25: Data from E. C. Hammond and D. Horn, 1966, appearing in John W. Kimball, *Man and Nature.* Copyright © 1975 Addison Wesley, Menlo Park, CA.

Fig. 16.26: Data from E. C. Hammond and D. Horn, 1966, appearing in John W. Kimball, *Man and Nature.* Copyright © 1975 Addison Wesley, Menlo Park, CA.

Fig. 28.5: Data from Population Reference Bureau.

Fig. 28.21: Data from Population Reference Bureau.

Fig. 32.7: Reprinted with permission from *Darwin and After Darwin* by G. J. Romanes. Copyright © Open Court Publishing Company, La Salle, IL.

Fig. 33.17: From Walter M. Fitch and Emanuel Margoliash, in *Science,* 155:279–284 (1967). Copyright © 1967 American Association for the Advancement of Science, Washington, DC. Reprinted by permission.

Fig. 37.24: From Katherine G. Field, et al., "Molecular Phylogeny of the Animal Kingdom" in *Science,* 239:748, 1988. Copyright © 1988 American Association for the Advancement of Science, Washington, DC. Reprinted by permission.

Fig. 38.20: From Charles Sibley and Jon Ahlquist, "The Phylogeny of the Hominoid Primates as Indicated by DNA-DNA Hybridization" in *Journal of Molecular Evolution,* Vol. 20, 1984. Copyright © 1984 Springer-Verlag/Publishers, New York. Reprinted by permission.

Photographs

Fig. 25.8: Photo courtesy of Carolina Biological Supply Company.

Fig. 33.20: From Biological Sciences Curriculum Study, *Biological Science: Molecules to Man,* Houghton Mifflin Co., 1963.